RECONFIGURING THE WORLD

JOHNS HOPKINS
INTRODUCTORY STUDIES
IN THE HISTORY
OF SCIENCE

Mott T. Greene
and Sharon Kingsland
*Series Editors*

Paul Lawrence Farber, *Finding Order in Nature: The Naturalist Tradition from Linnaeus to E. O. Wilson*

Anita Guerrini, *Experimenting with Humans and Animals: From Galen to Animal Rights*

Bruce J. Hunt, *Pursuing Power and Light: Technology and Physics from James Watt to Albert Einstein*

Trevor H. Levere, *Transforming Matter: A History of Chemistry from Alchemy to the Buckyball*

# Reconfiguring the World

*Nature, God, and Human Understanding*
*from the Middle Ages to Early Modern Europe*

Margaret J. Osler

THE JOHNS HOPKINS UNIVERSITY PRESS

BALTIMORE

© 2010 The Johns Hopkins University Press
All rights reserved. Published 2010
Printed in the United States of America on acid-free paper
9 8 7 6 5 4 3 2 1

The Johns Hopkins University Press
2715 North Charles Street
Baltimore, Maryland 21218-4363
www.press.jhu.edu

Library of Congress Cataloging-in-Publication Data
Osler, Margaret J., 1942–
    Reconfiguring the world : nature, god, and human understanding from the
Middle Ages to early modern Europe / Margaret J. Osler.
        p. cm.
    Includes bibliographical references and index.
    ISBN-13: 978-0-8018-9655-2 (hardcover : alk. paper)
    ISBN-10: 0-8018-9655-X (hardcover : alk. paper)
    ISBN-13: 978-0-8018-9656-9 (pbk. : alk. paper)
    ISBN-10: 0-8018-9656-8 (pbk. : alk. paper)
    1. Science, Medieval.  2. Science—Philosophy.  3. Science—Europe—
History.  I. Title.
    Q124.97.O85 2010
    509.4—dc22          2009048735

A catalog record for this book is available from the British Library.

*Special discounts are available for bulk purchases of this book. For more
information, please contact Special Sales at 410-516-6936 or specialsales@
press.jhu.edu.*

The Johns Hopkins University Press uses environmentally friendly book
materials, including recycled text paper that is composed of at least 30 percent
post-consumer waste, whenever possible. All of our book papers are acid-free,
and our jackets and covers are printed on paper with recycled content.

For Francine Michaud,
Scholarly companion and treasured friend

# Contents

Acknowledgments  ix

Introduction  1

1  The Western View of the World before 1500  3

2  Winds of Change: Searching for a New Philosophy
of Nature  30

3  Observing the Heavens: From Aristotelian Cosmology
to the Uniformity of Nature  61

4  Creating a New Philosophy of Nature  77

5  Shifting Boundaries: From Mixed Mathematics
to Mathematical Physics  94

6  Exploring the Properties of Matter: Alchemy
and Chemistry  118

7  Studying Life: Plants, Animals, and Humans  132

8  Rethinking the Universe: Newton on Gravity and God  147

Epilogue  165

Suggested Further Reading  169

Index  177

# Acknowledgments

Although all writing requires solitude, nevertheless it has taken a community of academic institutions and many helpful friends and colleagues to bring this book to completion. I am grateful to them all.

Several fellowships provided release time from teaching, giving me the opportunity to work full-time on this book. I am grateful to the University of Calgary for granting me sabbatical leaves for the fall term 2002, the winter term 2006, and the fall term 2009; to the Faculty of Social Sciences of the University of Calgary for a Research Fellowship for the fall term of 2006; to the Calgary Institute of the Humanities for a Fellowship during the academic year 2007–8; and to the Isaac Walton Killam Foundation for a Resident Fellowship for the winter term of 2009.

The Erskine Foundation at the University of Canterbury provided the opportunity for me to spend seven delightful weeks in Christchurch, New Zealand, as a Visiting Erskine Fellow during May and June 2008. During that time, I taught a six-week course on the scientific revolution to Philip Catton's History and Philosophy of Science 201 class, in which I used an early and incomplete draft of this book as one of the textbooks. I am grateful to both Philip and the students in the class for useful feedback on my draft and for many interesting discussions of related topics. Philip and Judith Catton generously shared their home with me, ensuring a warm human environment during that interlude.

A number of friends and colleagues took the time to read all or part of earlier drafts of this book and have provided useful comments and suggestions for improving it. Warm thanks go to Peter Barker, Ruth Barton, Paul Farber, Dorothy Grover, Wayne McCready, Ronald L. Numbers, Ellen Parker, Lawrence M. Principe, Stephen D. Snobelen, and Mark Waddell for their time and help.

Two anonymous readers for the Johns Hopkins University Press made insightful suggestions and detailed comments, which have led to extensive revisions and (I hope) improvements to the book. Mott Greene and Sharon

Kingsland, editors of the series, have been generous with their suggestions and encouragement. Josh Tong, at the Johns Hopkins University Press, has been a steady source of important information and assistance throughout this project. Deborah Bors, production editor with the Press, provided essential guidance during the final phase of this project, ensuring the professional quality of the finished book. I am especially grateful to Carolyn Moser for her thoughtful and sympathetic copyediting.

Robert J. Brugger, my editor at the Johns Hopkins University Press, has prodded and encouraged me through the slow process of bringing this project to completion and has been supremely understanding when the demands of life caused progress to grind to a halt. His blue pencil taught me important ways to improve my writing, lessons which will have a lasting effect. I have heard others compare Bob to that great literary editor Maxwell Perkins. The shoe fits.

My colleague Francine Michaud, to whom this book is dedicated, has given me invaluable support and encouragement through many years of friendship.

RECONFIGURING THE WORLD

# Introduction

Science and science-based technology dominate the twenty-first-century world, making it difficult to realize that other ways of understanding the world and other ways of doing things once prevailed. But, in fact, the worldview embedded in modern science is a relatively recent development in human history. This book aims to depict the ideas that prevailed at a time when educated people understood the world in ways very different from the one portrayed by modern science.

During the early modern era—roughly from 1500 to 1700—European thought underwent seismic shifts that shook the ground beneath most areas of intellectual activity: the arts, religion, philosophy, geography, medicine, and the sciences. The understanding of nature and the natural world lay at the foundation of these changes.

The roots of European intellectual life grew from the seeds planted by both the biblical religions and the ancient Greeks, a tradition frequently called the marriage between Athens and Jerusalem. This uneasy marriage was fraught with difficulties. The biblical account emphasized the unrestrained will of an all-powerful God, while the Greek approach emphasized a world ruled by impersonal principles of unity and harmony, usually without any moment of creation or interference by a willful God. Finding a workable relationship between these two outlooks permeates European intellectual history.

The Hebrew Bible, which Christians call the Old Testament, describes a world that God created and with which he continues to interact. He cares for his creatures, and he communicates with humans directly with rewards, punishments, prophecies, and miracles. The existence and nature of the world as well as the conditions of human life remain contingent on divine will. The Bible presents a history of the world and humanity, starting with the Creation and continuing through Noah's flood. The Christian New Testament continues the story, recounting the life, death, and resurrection of Jesus and concludes with prophecies about the Second Coming, the Final Judgment, and the end of the world.

Greek accounts of the world are very different. Unlike the biblical writings, they emphasize unchanging principles that provide the foundation and order of the natural world. Some philosophers considered these principles to consist of some kind of matter; others regarded them as mathematical or musical harmonies. Thus, where the Bible recounts nature and humankind as subject to the willful actions of an all-powerful God, the Greek philosophers focused on the regularity and order they perceived in the natural world. Moreover, the Greeks viewed the world as eternal, having neither beginning nor end.

The philosopher Aristotle (384–322 BC) exerted enormous influence in many areas of thinking about the natural world in his own day and through the Middle Ages and beyond. In addition to writing influential treatises on the physical and living worlds, he organized the branches of knowledge into categories defined by both their subject matter and their methods. His classification of the sciences provided the basis for university curricula from the thirteenth through the seventeenth centuries. The content of the various disciplines reflected his influence, as well, until traditional disciplinary boundaries began to erode in the sixteenth and seventeenth centuries.

Natural philosophy, which Aristotelians considered knowledge of the causes of all the phenomena in the world, was the disciplinary locus of much of this change. Although natural philosophy embraced many of the areas that we consider to be parts of science today, the term is not just an old-fashioned way of talking about science. Medieval natural philosophy included topics that modern science excludes, such as arguments for design in the world and the immortality of the human soul. And it excluded others, such as optics, astronomy, and medicine, which are important parts of modern science. Furthermore, natural philosophers sought a kind of knowledge—certain knowledge of the real essences of things—that differs in both method and goal from the epistemic ambitions of modern science.

How, then, did early modern thinkers regard nature and our capacity to know it? How did their ideas about these questions change in the early modern era? And what sorts of events, ideas, and traditions led to these changes?

This book aims to convey an understanding of how the natural world looked to early modern thinkers. What sorts of things did they think exist in this world? How did they explain these things and their changes? And how is such knowledge attained? How did the answers to these questions change between 1500 and 1700? And why?

The past is a foreign country. Rather than searching the past to find the origins of modern science, we aim here to find the early modern answers to these questions. Instead of discovering distorted reflections of our own preoccupations in the past, we seek to understand the language and customs of the inhabitants of this foreign land.

# 1 The Western View of the World before 1500

Knowledge of the world—the understanding of nature—develops by considering three fundamental questions: Of what sorts of things does the world consist? How do these things interact? And what kind of knowledge can we attain of them? On the journey from ancient Greece to sixteenth-century Europe, these questions will provide the guideposts for understanding how ideas about the natural world developed.

During the Middle Ages and Renaissance, scholars pursued questions about the world in institutions and books that embodied a particular classification of the fields of knowledge within well-defined disciplinary boundaries. During the early modern period, dramatic developments altered both the boundaries and the content of these disciplines. Around 1500, a university-educated person would think of the world in terms originally worked out in ancient Greece. In the two millennia between ancient Greece and sixteenth-century Europe, ideas about the natural world took a long and complicated journey. As scholars in other cultures appropriated Greek ideas for their own purposes, they modified them and put them to use in contexts different from the ones in which ancient thinkers had originally formulated them. To see the world in the eyes of educated Europeans around 1500 requires retracing some of these intellectual, geographical, and religious pathways.

Aristotle's natural philosophy provided the fundamental principles for explaining how the world works. The astronomy of Ptolemy (ca. AD 100–170) provided mathematical models for calculating the positions of the heavenly bodies, calculations useful for devising both calendars and astrological charts. The physicians Hippocrates (460–377 BC) and Claudius Galen of Pergamum (AD 129–216?) wrote treatises on the theory and practice of medicine that dominated medical thinking for centuries. These books and ways of thinking came to early modern Europe by linguistically and geographically circuitous routes, and both underwent major changes in the process. Aristotle, Ptolemy, Hippocrates, and Galen all wrote their books in Greek. Copied by hand, manuscripts of these works circulated in the Mediterranean world for more than a thousand years. The Roman conquest of the Greek world in the second century BC and the eventual decline of the Roman Empire in the fourth and fifth centuries of the current era resulted in the disappearance of the ancient centers of Greek learning—Athens, Rome, and Alexandria in Egypt.

After a long twilight, the rise of Islam in the seventh and eighth centuries stimulated a renewed interest in the learning of the Greeks. During the eighth and ninth centuries, scholars centered in Baghdad translated most of the works of Greek philosophy, natural philosophy, mathematics, astronomy, and medicine into Arabic. The translation movement received support from the caliphs (the chief civil and religious leaders, as successors of Muhammad), who believed that the Greek writings were originally part of the canon of Zoroastrianism (the chief religion of the Persian Empire) that came to be known among the Greeks as a result of Alexander the Great's pillage of Persia (modern-day Iran). The caliphs sponsored the translations as a way of recovering ancient Persian knowledge. In this way, the caliphs thought that they could convince the Persians that their newly founded dynasty was the legitimate heir to the ancient Persian Empire. In addition to philosophy, the Arabic translators paid special attention to astrology, a subject of particular significance to the rulers in Baghdad. A lively tradition of scholarship developed around these texts. In the wake of the translations, other Arabic scholars produced numerous new works in the sciences and philosophy. During centuries when the Latin West (Western Europe) was a politically decentralized, feudal society with only a limited tradition of learning and scholarship, philosophy, medicine, and the mathematical sciences flourished in the Arabic world.

Starting in the eleventh century, partly as a result of renewed contact with the Arabic world during the Crusades and the reconquest of the Iberian (Spanish) Peninsula by Christian Europeans, some European scholars undertook the retranslation of many of these Graeco-Arabic works from Arabic into Latin, then the scholar's language in Western Europe. By around 1200, most of the works of Aristotle and Ptolemy, as well as many other ancient writers, had been translated from Arabic into Latin. The availability of these texts and the excitement they created led to the founding of the universities, as groups of students engaged "masters" to teach them about the newly translated works. The University of Bologna developed around the Justinian Code, the basis of Roman law, the system of law that governed both the Church and civil society in continental Europe. In Paris, Peter Abelard (1079–1142) attracted students who wanted to learn Aristotelian logic, texts of which he possessed; the University of Paris, founded around 1200, thus became a center for the study of philosophy and theology—with an Aristotelian flavor. While not part of the Church per se, the University of Paris was closely associated with the Church in many ways. It was overseen by the bishop of Paris, and both students and faculty had clerical status: they were priests and monks. At Paris the Faculty of Theology—one of the three higher faculties, the others being medicine and law—dominated. Theology stood as the queen of the sciences in the medieval hierarchy of disciplines. The undergraduate, or arts faculty, officially had secular status—it was not concerned with religion or theology. The Aristotelian texts provided the base of its curriculum, even though Aristotle's philosophy was not always compatible with Christian theology. For example, he argued that the world is eternal, while according to Christian doctrine, God created the world, which thus had a beginning in time.

This lack of perfect fit produced a number of interesting consequences. Rather than rejecting Aristotle out of hand, a number of medieval scholars addressed the apparent conflict between Christian doctrine and Greek philosophy by seeking ways of making them compatible. The most influential of these attempts was the great synthesis, the *Summa theologica* by the Dominican friar Thomas Aquinas (1224/27–1274). In this massive work, Thomas attempted to explain Christian doctrine by means of Aristotelian philosophy. Because of some of the contradictions between Aristotle's views and Christian doctrine— such as the absence of a personal or providential God in his philosophy or his denial of a human soul that could be separated from the body—Thomas modified many of the ancient philosopher's claims. The new university and its members changed from transmitters of ancient knowledge to creators of new knowledge. As a consequence of works like the *Summa theologica* as well as the organization of the university curriculum, medieval Aristotelianism and Christian theology became closely intertwined in a philosophy called Scholasticism. Scholastic thinkers and the Scholastic curriculum continued to dominate most European universities until the seventeenth century and beyond. Then, as now, scholars faced decisions about what to pass on and what to change, yet always within the bounds of certain texts, such as Aristotle's.

## Understanding the World: Aristotelian Natural Philosophy

From the time the works of Aristotle were translated into Latin (ca. 1200) until the seventeenth century AD, almost 400 years later, most European thinkers concerned with the nature of the world accepted some version or another of the Aristotle's account of nature and knowledge. Aristotle came from the small city of Stagira in Macedonia—a harsh rural land in the north of Greece. He studied philosophy in Plato's Academy in Athens, where he spent twenty years learning and teaching. Later, he served as tutor to the young man who would become Alexander the Great. He also spent several years on the Greek island

Lesbos, where he made extensive empirical studies of marine animals and wrote about land animals as well. Later Aristotle established his own philosophical school, the Lyceum, in Athens, and most of the writings now attributed to him are probably based on lectures he gave at this school. In what became known as his "natural books"—*The Physics, On the Heavens, On Generation and Corruption, Meteorology, On the Soul, The History of Animals, The Parts of Animals,* and *Generation of Animals*—Aristotle defined the conceptual framework in terms that were used for discussions and explanations of nature and the natural world for the next two thousand years. Thus, time spent understanding Aristotle's views is time well spent.

Aristotle began his discussion of the natural world by describing the different areas of knowledge. His classification of knowledge defined the disciplinary boundaries that would later govern most treatises, textbooks, and university curricula. Aristotle distinguished physics from both mathematics and theology on the basis of its subject matter: Physics deals with substance that is perishable and sensible; mathematics deals with substance that is sensible but not perishable; and theology deals with substance that is not sensible and not perishable. Because of the differences between the subject matter and the principles of each of these disciplines, Aristotle concluded that they are distinct and unrelated. That he considered it illegitimate to apply mathematics to physics became particularly relevant for later developments in natural philosophy and the sciences. This approach may sound odd to us, but it made perfect sense to our predecessors.

Aristotle believed that in any area of inquiry we attain understanding through knowledge of its principles, causes, or elements. He maintained that the world consists of substances—meaning combinations of matter and form. Individual substances are things such as cats, humans, stones, and statues. Each substance is composed of two components, matter and form: the matter is the stuff out of which it is composed, and the form is that which makes it what it is rather than something else. A cat consists of matter (its flesh, and, fur, and bones) and its form, that which makes it a cat rather than a dog or a statue. For Aristotle, each individual thing has a nature, its inner cause of motion. For him the word "nature" refers to the essence of an individual (as in "It is the nature of gamblers to take risks"), not to the world as a whole or to the part of the world untouched by humans; these meanings emerged only in later centuries. Aristotle, like other ancient Greek thinkers, defined physics or natural philosophy as the study of the unique natures of things taken separately, not Nature, taken all together.

Forms in Aristotle's world can be either actual or potential: what they are or what they can become. A mature oak tree actually possesses the form of the oak. The acorn, however, contains the form only potentially. Potentiality becomes actualized, to use Aristotelian jargon, when the acorn grows into an oak. The process of actualizing the form controls the development of the tree. Because forms control development, acorns produce oaks but not maples, and cats produce kittens but not puppies. Forms govern every natural process. This relationship between potentiality and actuality brings us to the question of how things change.

According to Aristotle, several different kinds of change occur in the world: change of place or local motion; quantitative change, such as expansion and contraction; qualitative change, such as the reddening of an apple as it ripens; and coming-to-be and passing away, such as birth and death. Every change requires its own causal explanation. For Aristotle, a complete explanation appeals to four kinds of causes: the formal cause, or form; the material cause, or the matter; the efficient cause, or the agent that brings about the change; and the final cause, or goal or purpose of the change.

The construction of a house illustrates the nature of the four Aristotelian causes. The architect's plan, the blueprint, is the formal cause. The matter used to construct the house—the bricks and mortar, the two-by-fours, pipes, wires, roofing material, and drywall—is the material cause. The activity of the workmen who construct the house is the efficient cause. And the final cause or purpose of the house is to provide shelter. The four causes also explain changes in the natural world—in the absence of an external or a conscious agent. Consider the growth of an oak tree: the formal cause is the actualization of the form of the oak, which exists potentially in the acorn; the material cause is the water, earth, and other matter of which the tree consists; the efficient cause is the "actualization" of the form from potentiality to actual oak tree; and the final cause is the production of an offspring that resembles its parents, that is, the actualization of the form of oak. In examples of living things like the development of an oak from an acorn, the formal, final, and efficient causes are often the same.

In this system of thinking, every natural change—that is, every change that the inner nature of the thing produces—has a final cause or end. The Aristotelian world is thus profoundly teleological or goal-directed. Not all changes are natural. Those that lack final causes—that is, built-in goals or aims—are coincidental in the sense that the outcome of the process does not result from goal-directed change. Accidents occur in several ways. Sometimes an otherwise natural process does not reach its proper goal. For example, while the reproductive process normally results in the birth of an offspring resembling its parents, if something goes wrong during conception or gestation, a miscarriage or a monstrous birth results and lacks a final cause. The deformed offspring results from some accident such as an injury to the pregnant woman or a genetic anomaly that prevents the production of an offspring resembling its parents. Likewise, conscious agents act for ends, but not all their actions are purposeful. For example, a teacher of grammar aims to teach the proper use of language. If the teacher makes a grammatical error, however, that error does not have a final cause and is simply an accident.

Aristotle incorporated the entire range of natural phenomena into his conceptual framework. In his treatise *On the Heavens*, for example, he dealt with cosmology, using his concepts of matter, form, and the four causes to describe the celestial region above us and to contrast it with the terrestrial region that we inhabit. The Aristotelian universe, or cosmos, is spherical. Its center coincides with the center of the earth, and its circumference is a sphere lying outside of the sphere of the fixed stars. The Greeks thought that the planets are stars that move, or "wandering stars." Thus what we call "stars" they called "fixed stars." According to Aristotle, the cosmos is divided into two regions, which consist of different kinds of matter. The region inside the sphere formed by the moon's orbit—the sublunar or terrestrial region—consists of the four elements: earth, water, air, and fire. The region beyond the sphere of the moon—the heavens or celestial region—consists of a fifth element, quintessence, which is arranged in a set of concentric spheres. Bodies in the sublunar region undergo qualitative change and tend to move in straight lines. The heavenly bodies—the sun, the moon, and the five planets (Mercury, Venus, Mars, Jupiter, and Saturn)—do not change, although they move from place to place. Spheres of quintessence hold them in place, and all the heavenly bodies revolve around the earth, which is at the center. The cosmos is full of matter: void or vacuum does not exist. This picture is quite different from our understanding, but not from our experience: it is partly, at least, what we see. In the cosmos, as elsewhere, local motion (or change of place), like all change, always requires a cause. The bodies in the celestial region, above the sphere of the moon, consist of the fifth element, quintessence, and so they naturally move in circular motions around the center of the cosmos. The nature of quintessence causes their motion.

Different, more complicated motions characterize the terrestrial region. Downward and upward motions, caused respectively by the heaviness or lightness of the moving body, are called "natural" motions. Heavy bodies naturally move downward because the form of heaviness causes them to seek their natural place at the center of the cosmos. Light bodies move upwards because the form of lightness causes them to seek their natural place—upward, which is the circumference of the region lying within the sphere containing the moon's orbit. Motions that have external causes, such as projectile motion or the motion of lifting a heavy stone, are called violent. Based on these distinctions, Aristotelians developed a complex scheme for explaining the causes of all kinds of motions. Only a body at rest does not require explanation in this scheme. From that perspective, the Aristotelian world is fundamentally static: it creates a special and important role for being at rest.

In his treatise *On Generation and Corruption*, Aristotle addressed basic questions about matter and change. Matter, in the Aristotelian world, always exists in combination with form. Matter is informed, and forms are embodied. Prime matter, an abstraction which we never actually see by itself, becomes informed by combinations of the four primary qualities—the hot, the cold, the dry, and the moist—to produce the simplest level of matter, the four elements. Fire is hot and dry; air is moist and hot; water is cold and moist; and earth is cold and dry. All other kinds of matter consist of various proportions of these elements. Unlike modern chemical elements (except for radioactive ones, whose decay process was unknown until the twentieth century), Aristotelian elements can change one into another. For example, if the hotness of fire becomes cold, fire can be transformed into earth, or if the coldness of water becomes hot, water can be transformed into air. Matter can be informed at various levels. The bronze matter of the statue of the horse has the form of horse. But considered as a chunk of metal, it has the form of bronze. That we do not think in these terms is not as important as understanding that for a very, very long time, most natural philosophers did.

In *Meteorology*, Aristotle set out to explain various phenomena that he believed exist or occur in the earth's atmosphere, including shooting stars, comets, the Milky Way, various kinds of precipitation, and weather, as well as the saltiness of the sea, earthquakes, thunder and lightning, rainbows, and a host of other things. His curiosity in this research was quite like our own even if his explanations differ from ours.

Included among his natural books is one entitled *On the Soul*. Aristotle claimed that three kinds of soul exist, corresponding to the three kinds of living things. Plants, animals, and humans all possess a vegetative soul, the form that endows the organisms with capacities for nutrition, growth, and reproduction. Animals and humans, but not plants, possess in addition a sensitive soul that endows the organisms with sensation and the ability to move around. Only humans possess a rational soul, giving them the capacity for rational thought. The soul is the form of the organism and passes from one generation to the next by means of the process of reproduction, a process that he described and explained in his book *Generation of Animals*.

In addition to his account of the soul, Aristotle wrote several descriptive books on the appearance and behavior of animals, which he based on his extensive empirical researches. Aristotle explained animals using the same principles that he used to explain every other kind of natural thing. In fact, it is likely that his observational study of living things provided the foundation for his philosophy more generally. That connection would explain both his conception of the embodied characteristic of forms and the importance of final causes and ends in explaining change.

Taken collectively, Aristotle's natural books bring the entire world within the compass of his unified philosophy of nature. He applied the concepts of matter, form and the four causes to explain all natural phenomena. These general features of Aristotle's works continued to characterize subsequent works on natural philosophy down to the seventeenth century.

In addition to describing the general principles according to which the world functions, Aristotle described a method for acquiring knowledge of the world. Aristotelian natural philosophers sought demonstrative knowledge of the essences of things. Demonstration of the properties of things from knowledge of their essences or forms leads to this goal. Aristotle spelled out a method for discovering the essences of things by examining a number of individuals of the same kind to determine which properties are essential and which are merely accidental. For example, after examining a number of humans, Aristotle concluded that reason, absence of feathers, and two-footedness are essential qualities of all humans, while white skin, a snub nose, and curly hair are not. Once the philosopher identifies the essence of a kind of being, he can demonstrate its other properties. Complete knowledge of a thing involves not only knowledge of its form, but also an explanation based on the four causes. Once complete, such knowledge achieves certainty.

Aristotle's philosophy of nature thus provided a complete account of the world. Using his fundamental ideas of matter, form, and the four causes, he described and explained the motions of the heavens, the physical properties of material things on or near the earth, the animate world, and human beings. These ideas also formed the basis of his other writings on ethics, political philosophy, and even literary criticism. The coherence and unity of his philosophy accounts, at least in part, for its attractiveness to so many thinkers for many centuries. Continued interest in Aristotelianism and the further development of theories within a fundamentally Aristotelian framework followed a long and winding road, passing through diverse linguistic and cultural contexts.

In the early Middle Ages, scholars in the Latin West knew only a few of Aristotle's works. At the same time in the Arab world, however, scholars translated virtually the entire body of his writings from Greek into Arabic. By the end of the ninth century, philosophers writing in Arabic undertook to develop their own interpretations of Aristotelian philosophy. Among these, Abū Ysuf Ya'qub ibn Ishaq al-Kindī—known simply as al-Kindī (d. 866–73)—oversaw the work of the translators and developed a new philosophical language in Arabic to handle Aristotle's ideas. He argued that these philosophical ideas pertained to solving the problems of his own time, including problems emerging from Islamic theology. He wrote numerous treatises on the sciences as well as completing what he thought the ancient philosophers had left unfinished. He referred to Aristotle as "The Philosopher," a designation that remained throughout the Arabic philosophical tradition. Thus we see that, as in the later case of Thomas Aquinas, transmission and creation went together.

The Spanish Muslim Abū al Walid ibn Rushd, or Averroes (1126–98), matured in a religious tradition that emphasized the rationality of human understanding and the unity of God and his creation, as well as the rationality of the *Qur'ān* and its interpretations. In this context, he created ideas in philosophy and theology based on natural reason as opposed to revelation. His writings on Aristotle created his reputation as "The Commentator." Averroes emphasized the unity of truth. When apparent contradictions arose between rationally demonstrated philosophical truths and Scripture, he believed that the contradictions could be resolved by interpreting the scriptural passages allegorically. Some of his resulting claims proved to be extremely controversial in the Christian environment of thirteenth-century Paris, notably his denial of creation *ex nihilo* (the absolute creation of something from nothing), his claim that divine knowledge is qualitatively different from human knowledge, his denial of the survival of the individual after death, and his deterministic interpretation of Aristotelian natural philosophy (that is, an interpretation of causes acting with inexorable necessity). For example, Averroes claimed that there is an unchangeable set of necessary causes linking the motions of the heavens to events in the terrestrial world, including human activities. Causal necessity of this kind would limit both divine and human free will, a view that is unacceptable to Christians.

The translation of Aristotle's works from Arabic into Latin stimulated efforts to use the ancient philosophy as a foundation for Christian theology.

Thomas Aquinas wrote systematic treatises in which he deployed Aristotelian philosophy to explain matters of Christian doctrine. For example, he used Aristotelian matter theory to explain how Christ can be really present in the elements of the mass even though the bread and wine do not change their appearance. He claimed that only the substantial form changes from bread and wine to body and blood of Christ, but the appearance of the elements remains the same. Herein lies the miracle of the Eucharist. The Church adopted this explanation, called transubstantiation, as official dogma. Because many Christian doctrines received Aristotelian explanations, natural philosophy became inextricably linked to religious doctrine.

Aquinas asserted that natural reason cannot prove every doctrine, notably the mysteries of creation, resurrection, and the Trinity. Truths known by faith, he claimed, do not contradict those known by reason, but complete them. Thomas' teacher, also a monk in the Dominican order, Albertus Magnus, or Albert the Great (1200–1280), claimed that natural reason is autonomous, and he wrote treatises on various topics of natural history.

At the same time, Averroes' ideas gained currency at the University of Paris in the teachings of Siger of Brabant (ca. 1240–84) and other Latin followers of The Commentator, leading to the Condemnation of 1277, in which the bishop of Paris condemned 219 propositions, many of which denied God's free will. Ironically, the Condemnation stimulated philosophers to develop new ways of thinking about God's relationship to the creation, ways which emphasized divine freedom. Some of these ideas resulted in a reconsideration of traditional views about the nature and structure of the world. For example, one of the condemned propositions stated that the earth must be stationary at the center of the cosmos. No one at the time denied that the earth is at the center of the cosmos and that it is stationary, but, given God's omnipotence and freedom, it should be possible for him to have created a universe in which the earth moves and is not at the center. Two fourteenth-century natural philosophers—John Buridan (ca. 1300–1358) and Nicole Oresme (ca. 1325–82)—considered this possibility at length. They presented answers to the traditional objections to the earth's motion to demonstrate its possibility but concluded that in fact the earth is stationary at the center of the cosmos.

The study of natural philosophy continued to be based on Aristotle's *libri naturales*, or natural books, for the next three centuries, until around 1600. Although scholars in the Renaissance produced new editions of Aristotelian texts more faithful to the originals than those that were available during the Middle Ages, the use of these improved texts did not immediately alter the Aristotelian structure of the study of natural philosophy. Textbooks on natural philosophy continued to follow the structure and order of Aristotle's natural books. Even when the content of natural philosophy underwent enormous changes in early modern times, university curricula and textbooks continued to follow the outward pattern of their medieval predecessors. This is an important point: there is a tendency to exaggerate the rupture between medieval and Renaissance natural philosophy, but in fact many traditional forms endured.

## Saving the Phenomena: Ptolemaic Astronomy

Greek astronomy developed as a mathematical discipline quite apart from natural philosophy. It had a long history reaching far into the past. In Greek thought, unlike modern science, calculation was a technique for predicting the positions of heavenly bodies, not a source of truth about the cosmos.

Mathematical astronomy has ancient roots. The Babylonians had observed the heavens systematically from at least as early as 1700 BC. They recorded their observations on clay tablets, hundreds of thousands of which now reside in the British Museum in London. Using arithmetic functions, the Babylonians were able to make approximate predictions of the positions of the heavenly bodies relevant to their needs for a calendar and for judicial astrology (predicting future events or establishing propitious times for marriages, founding a city, or other ceremonial events on the basis of the configuration of the heavens). As far as we know, they never used geometrical models for their astronomical calculations. Instead they used columns of numbers, some recording their meticulous observations of numerous variables and some resulting from arithmetic manipulation, to predict such phenomena as eclipses and the first risings of the new moon (which determined the beginning of their lunar month). It was an astronomy of numbers and time, not objects and shapes in space.

By the third century BC Greek astronomers had access to much of the Babylonian astronomical data. But Greek methods of observation and calculation differed substantially from those of their Babylonian predecessors. A comment that the philosopher Plato (427–347 BC) supposedly made to the astronomers at his academy in Athens provided the stimulus for the methods the Greek astronomers adopted. For Plato, only reason—abstract reason not corrupted by the illusions of the senses—can produce reliable, certain knowledge. Because he distrusted the senses as a source of reliable knowledge and because astronomy necessarily relies on observational data, Plato instructed the astronomers to "save the phenomena" using only combinations of uniform circular motion. That is to say, he instructed the astronomers to construct theories which were consistent with all observed data and which could make accurate predictions about the future positions of the heavenly bodies, but also

## The Celestial Sphere

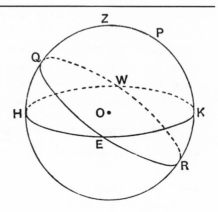

The Greek astronomers pictured the celestial sphere as the frame of reference for their geometrical theories of the motions of the heavenly bodies. The circle *HZKR* is the sphere of the heavens. The great circle *HWKE* is a projection of the earth's equator onto the celestial sphere. It is called the equator. The great circle *RWQE* is the annual path of the sun around the earth. It is called the ecliptic and lies at an angle of 23½ degrees to the equator. The points *W* and *E*, where the ecliptic crosses the equator, are the vernal and autumnal equinoxes. The points *Q* and *R* are the summer and winter solstices. Greek astronomers used the ecliptic as the east-west axis for measuring the longitude—the angular motion of the planets, the sun, and the moon in relation to the vernal equinox. The angu-

lar distance north or south of the ecliptic determined the body's latitude.

■ Reprinted from Arthur Berry, *A Short History of Astronomy from Earliest Times through the Nineteenth Century* (London: John Murray, 1898), p. 38.

---

instructed them that they should never consider their theories to be descriptions of the real structure of the heavens. In his influential dialogue *Timaeus*, in which he described the origin and structure of the cosmos, Plato stated that such an account can at best be a likely story but cannot be known with the certainty that characterizes genuine (purely rational) knowledge.

Thus Aristotelian physics provided a real description of the world's structure, but astronomy served merely as a device for calculating the positions of the heavenly bodies. Most Greek philosophers and astronomers did not find the methodological inconsistency between astronomy and physics problematic. If we cast this approach in terms of Aristotle's classification of the disciplines, they considered astronomy to be a branch of mixed mathematics because it used mathematics to describe changeable, observable phenomena. One of the most significant changes that occurred in late medieval and early modern times was a revision of the relationship between these disciplines.

Greek mathematical astronomy emerged in the fourth century BC. It aimed to account for the motions of the known planets (Mercury, Venus, Mars, Jupiter, and Saturn), the sun, and the moon—all of which are visible with the naked eye—by means of models constructed from combinations of uniform circular motions. Each of these bodies rises in the east and sets in the west daily, although the planets are not visible at all times of the year from a single place on earth. The annual path of the sun from west to east was called the ecliptic—a circle which the planets seem to follow—and served as the horizontal coordinate for measuring positions in longitude of the other heavenly bodies. It is not necessary here to master these technical details so much as to appreciate that they existed and, more to the point, that they were the basis of much theory and observation.

In addition to rising and setting daily, each body moves more slowly from west to east along the ecliptic over longer periods of time. Their periods—the time it takes each of them to return to the same position in relation to the fixed stars—range from about a month for the moon to about twenty-nine years for Saturn. In addition to periodic motion along the ecliptic, each planet appears to slow down, stop, and move from east to west, creating a loop in its generally west-to-east motion. These stops and reversals are called stations and retrogradations. The Greek astronomers accounted for the observed motions of the planets by using combinations of nested spheres and circular motions in creative ways, but further observations kept undoing their work.

Hipparchus (middle of the second century BC) had access to Babylonian observations that went back to at least 1700 BC as well as to the geometric tradition of his Greek predecessors. Utilizing about 1500 years of recorded observations in conjunction with important advances in mathematics, he developed theories that were more accurate than any that had been worked out by his predecessors. In order to retain a geocentric model for the planets, he used a new geometrical device, the eccentric circle. He also developed a model based on the epicycle, which the mathematician Apollonius of Perga (ca. 262–190 BC) had proven to be mathematically equivalent to the eccentric: either model could be used to "save the phenomena"—that is to construct a mathematical theory that was consistent with observed data and generated true predictions with the same degree of accuracy. His emphasis on the equivalence of the eccentric and epicycle underscores that he did not intend his models to describe physical reality.

Claudius Ptolemy gave a complete account of the heavenly bodies. His book the *Megale syntaxis* (*Great Collection*)—better known by its Arabic name, the *Almagest*, after translations into Arabic and later into Latin—served as the starting point for medieval and early modern developments in astronomy, reminding us again of the central role of Arabic scholars. Ptolemy used the epicycle to construct theories of the sun, the moon, and the planets. The *Alma-*

## Hipparchus' Demonstration of the Equivalence of the Eccentric and the Epicycle

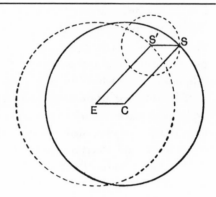

This diagram illustrates the equivalence of the epicycle and the eccentric circle in describing the motions of the planet S. S moves on the epicycle, which has a radius SS' that equals the eccentricity of the eccentric circle which has the earth at its center. The planet S moves at uniform angular velocity around C, but has an irregular velocity when viewed from E, the earth. It traces out the same path if it moves on the epicycle, which in turn moves uniformly around the deferent. E is at the center of the deferent.

■ Reprinted from Arthur Berry, *A Short History of Astronomy from Earliest Times through the Nineteenth Century* (London: John Murray, 1898), p. 47.

*gest* deals with each planet individually. It does not outline a comprehensive system, since Ptolemy's aim was to solve the problem of each planet's motion individually, not to produce a descriptive account of the cosmos. His theories predicted planetary positions accurately to within about 10 minutes of arc, the limit of accuracy of the observations available to him. (The sky contains 360 degrees of 60 minutes each, or 21,600 minutes, so 10 minutes of arc is a division of the visible hemisphere of the earth into 1,000 parts, or an arc of 1 percent.)

Using Aristotelian principles, Ptolemy argued that the earth is spherical, a fact that the Greeks knew and accepted even before Aristotle. Ptolemy asserted that the earth is made of heavy matter. Consequently, if the earth were removed from its natural place, it would tend to fall towards the center, its natural place. Since it is heavier than animals and other objects that are not on its surface, they would be left behind if the earth were to fall. But we observe no such effects. Moreover, since light things—like the fiery stars—are naturally more inclined to undergo rapid motion than heavy objects like the earth, the motion of the heavenly spheres is a simpler hypothesis than the rotation of the earth on its axis to account for the daily risings and settings of the planets, the sun, and the moon. For these reasons, Ptolemy argued that the earth is stationary at the center of the cosmos.

Ptolemy did not stop here, however. In addition to the *Almagest*, which

## Ptolemy's Epicycle

The planet *J* moves at a uniform angular velocity around the epicycle, the small circle which has its center on the larger circle, called the deferent. By adjusting the speeds and directions of the motions along the two circles, it is possible to generate any closed curve and thus account for the observed motions of the planet, including the stations and retrogradations. By tilting the plane of the epicycle in relation to the plane of the deferent, one can account for the planet's motion in latitude—that is, its varying distance from the ecliptic.

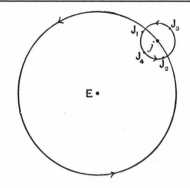

■ Reprinted from Arthur Berry, *A Short History of Astronomy from Earliest Times through the Nineteenth Century* (London: John Murray, 1898), p. 70.

deals with mathematical astronomy, Ptolemy wrote another book, *Planetary Hypotheses*, in which he attempted to give a physical account of the universe that differed from Aristotelian cosmology. Medieval Arabic astronomers knew this work, and they considered the differences between the mathematical astronomy of the *Almagest* and the attempted physical account in the *Planetary Hypotheses* to be highly problematic.

Ptolemy considered astrology, a subject that first appeared in Greek writings around the second century BC, to be an important part of astronomy. He gave a detailed account of it in the *Tetrabiblos* (*Four Books*), which opens with a discussion of the relationship between astronomy and astrology. Here, he explained that the kind of prediction that he had treated in the *Almagest* is fundamental, and that a second kind, astrology, depends on the first for the accurate data required for the construction of horoscopes. He described and explained the way heavenly bodies affect terrestrial events, from the effects of the sun and moon on the weather and the growth of plants to the effects of the heavenly bodies on the lives of individual people. He believed that the interconnections of all things in the cosmos are the causes of astrological influences. Because he thought that the positions of the heavenly bodies at the time of birth affect individual temperaments, he believed that astrology could play an important role in medicine. Astrology and astronomy remained closely connected through the Middle Ages and into the early modern period. That many thinkers today reject astrology cannot blind us to the strong impulse it provided for fundamental research in astronomy.

Although intellectual life waned in Europe during the early centuries of the first millennium AD, Arabic astronomers in Baghdad and other Middle Eastern centers became acquainted with Greek astronomy even before the rise of Islam in the seventh century. During the eighth and ninth centuries, they translated almost all serious Greek works in the sciences and philosophy into Arabic. Arabic astronomers wrote extensive commentaries on Ptolemy's *Almagest*. They severed Ptolemy's mathematical astronomy from astrology, which they rejected as contrary to religion because the deterministic influence (the causal necessity) of the heavenly bodies seemed to restrict both human and divine freedom. Their examination of his astronomy led to serious criticisms of his work and to the development of new mathematical methods that eventually influenced astronomy in early modern Europe. Again, transmission, translation, and transformation traveled together.

Seven centuries had passed between the time Ptolemy wrote the *Almagest* and the time when Arabic astronomy got under way. Over that long interval, small initial observational errors became hugely magnified. To correct these errors Arabic astronomers made new observations, using newly invented instruments. The new observations and the development of new trigonometric functions, an extensive improvement of Ptolemy's already very impressive work in the calculation of angles, became the basis of their revisions to Ptolemy's mathematical constructions.

A major problem with Ptolemy's astronomy was the same problem that Ptolemy's Greek successors had been unable to solve: the incompatibility of the geometrical constructions in the *Almagest* with the physical system of *Planetary Hypotheses*, in which Ptolemy had used only uniform circular motions. Also observation and theory were at odds. In the *Almagest*, Ptolemy had introduced a mathematical device called the equant, which seemed to "fudge" the commitment to perfect circles. He used the equant to capture some of the observed irregularities of planetary motions.

By the eleventh century, most Arabic astronomers regarded the equant as absurd because it implied that a physical sphere could move uniformly in place around an axis that did not pass through the center of the circle. This characteristic of the geometrical models rendered them incompatible with the physical reality of the celestial spheres. Every model of planetary motion that Ptolemy used in the *Almagest*—except for the sun—employed an equant. Ibn al-Haythan, known in Latin as Alhazen (d. ca. 1040), focused on this problem. He proposed new principles for astronomy, based on the idea that physical

## Ptolemy's Equant

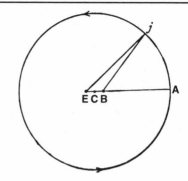

Ptolemy introduced the equant to account for the irregular angular velocities of the planets, sun, and moon around their orbits. The planet $J$ moves on the earth-centered circle but at an irregular rate. It moves uniformly around the point $B$ (the angle $JBA$ increases uniformly). The earth is not at the center. $B$ is at the same distance from the center, $C$, as is the earth, $E$, but on the opposite side of $C$. Viewed from $E$, the motion of $J$ is not uniform. Ptolemy used this device to capture some of the irregularities observed in the motions of the planets. Many later astronomers thought that Ptolemy's use of the equant violated the principle of uniform circular motion.

■ Reprinted from Arthur Berry, *A Short History of Astronomy from Earliest Times through the Nineteenth Century* (London: John Murray, 1898), p. 71.

objects should be represented mathematically by models that did not contradict their physical natures. Essentially, he regarded mathematics as a language for describing physical reality rather than just a set of constructions to save the phenomena. His approach amounted to a call for a new astronomy. Whereas the Greeks had used mathematical constructions simply as ways of predicting the positions of the heavenly bodies, more or less the way we would use an electronic calculator to solve complicated arithmetic problems, Alhazen insisted that the mathematical models of planetary motions actually describe the real motions of those bodies.

This work continued for at least two more centuries. In the thirteenth century, Mu'ayyad al'Dīn al 'Urdī (d. 1266) and Naṣr al Dīn al-Tūsī (d. 1274)—both of whom worked at the observatory at Marāgha founded by the Mongols in Persia in 1259 as an important center for both astronomical observation and scholarship more generally—proved previously unknown geometrical theorems that enabled astronomers to describe irregular motions of the heavenly bodies without violating the cosmological principle of uniform circular motion. A mathematical device that later became known as the Tūsī-couple enabled al-Tūsī to account for the oscillations of the sun and other planets in latitude without violating the principle of uniform circular motion. The Arabic astronomers continued to develop Alhazen's transformation of the role of mathematics, treating it as a language for describing physical phenomena. A century later, Ibn a-Shāṭir (d. 1375), who came from Damascus, insisted on matching mathematical models to the observations, thus redefining astronomy as a discipline that produces a systematic and accurate description of the real, physical universe.

Astronomers in sixteenth-century Europe became aware of these new mathematical theorems and the revised status of both mathematics and astronomy, probably by contact with European scholars who could read Arabic. Significantly, Arabic scholars did not simply "pass the torch" of Greek astronomy to Europe; rather, they had through their own work changed the nature and disciplinary status of astronomy in fundamental ways.

### Transforming Matter: Ancient and Medieval Alchemy

The earliest knowledge of the properties of different kinds of matter came from technologies like brewing, ceramics, and metallurgy. Mining, smelting, and the production of the alloy bronze became common, first in Mesopotamia and Egypt, and then throughout the entire Mediterranean region during the period between 4000 and 2000 BC. Although we have no record of any theories explaining the processes used, these technologies generated a considerable amount of experiential knowledge about the properties of various kinds of matter, particularly metals.

The early Greek philosophers developed theories of matter as part of their attempts to explain the natural world. Aristotle and the atomists had the greatest impact on the history of chemistry. Aristotle stated that all material substances consist of various proportions of the four elements—earth, water, air, and fire. Each of the elements could change into any of the others. In addition, Aristotle claimed that there is a limit to the size of particles that can be informed by the primary qualities (hot and cold, dry and moist). He called these smallest particles *minima*. Below that size, they would revert to prime matter. Thus, some Aristotelian texts described matter as having a quasi-particulate nature, although Aristotle and his followers rejected the atomism of Democritus and Epicurus. Aristotle also thought that the underground action of heat on exhalations from the interior of the earth forms metals.

The atomists claimed that all material things are composed of microscopic atoms that move in empty space. They thought that the atoms are indivisible and that they are all composed of the same kind of matter. Only the shapes of atoms differ. The shapes of atoms, the configuration of the clusters formed by collections of them, and their motions produce all the qualities—including the chemical properties—that we perceive in macroscopic objects. Atomists explained chemical changes as the reshuffling of the atoms that compose substances.

Chemistry and alchemy originated in Hellenistic Egypt, where the city of Alexandria, established as the capital of the Greek colony of Egypt by Alexander the Great (d. 331 BC), became an important center of learning and research in the sciences. Manuscripts from the third century AD contain recipes for a variety of chemical and metallurgical processes, including recipes for making alloys that look like gold and silver and for making artificial gems. The origins of alchemy itself are shrouded in mystery and legend. The named authors of many alchemical texts may not even have existed. The Alexandrian writers—prominent among whom were pseudo-Democritus (ca. AD 200), Zosimos of Panopolis (fl. AD 300), and the legendary Maria the Jewess (also known as Maria Prophetissa, reputed to be Miriam, sister of Moses)—contributed a number of procedures and laboratory apparatus that continued to play important roles in later alchemy and chemistry. Techniques such as distillation, sublimation, and fixation remained central to alchemical practice. Some of the writers, most notably Zosimos, described chemical processes in an allegorical style, a feature that became common in later alchemical writing.

In addition to describing techniques for making imitations of gold and silver, the Alexandrian writers began to describe procedures for actually making gold and silver from base metals such as lead and copper. The manuscripts instructed the practitioner to start by making a substance, usually a powder, called the "philosophers' stone." The search for the philosophers' stone became the central goal of alchemists through the seventeenth century.

When Arabic scholars translated the bulk of Greek science and philosophy into Arabic in the eighth and ninth centuries, they included a wide variety of Greek alchemical texts. In fact, the word "alchemy" derives from the Arabic word *al-kīmiyā*, which is composed of the Arabic definite article *al* and the Greek word for chemistry, *chymia*. As with other areas of knowledge, the Arabic scholars wrote original works on alchemy as well as translating the Greek works and writing commentaries on them. The two most prominent writers on alchemy were Jābir ibn-Hayyān (d. 812) and Muhammad ibn-Zakarīyā ar-Rāzī (ca. 865–ca. 925), known in the West as Rhazes. Jabir probably did not write all of the 2,000 books ascribed to him; later authors likely attributed their own writings to him. In fact, his name may not even refer to a real person.

The Jabirian books rely on Aristotle's theory of the elements, treating them as concrete substances that can be isolated from more complex bodies. According to this theory, each substance in the world consists of definite proportions of each of these elements. The books contain descriptions of precise, quantitative methods to calculate the proportion of the elements in any given kind of matter. Various chemical processes can extract the elements composing material substances and then reassemble them to form other kinds of substances. This theory of matter provided one theoretical foundation for the belief in the transmutation of metals.

A different Jabirian book stated that metals are composed of mercury and sulfur. Based on Aristotle's theory of underground exhalations, this theory claimed that the two underground exhalations condense, so that the moist exhalation becomes mercury and the dry one becomes sulfur. The resulting mercury and sulfur then combine to form the various metals, their differences caused by different proportions of the constituent mercury and sulfur. When impurities contaminate the mix, the other, baser metals are produced. Variations of this theory survived until the eighteenth century.

Rhazes, who ran hospitals in the Persian city of Rayy and later in Baghdad, made his reputation largely as a medical writer. In addition to books on medicine, he wrote several important alchemical treatises that surveyed the field of alchemical knowledge, classifying substances and describing equipment and techniques. He adopted the mercury/sulfur theory of metals and discussed the transmutation of metals. 'Alī Abū al-Husayn Ibn 'Abdallāh ibn Sīnā (980–1037), known in the Latin world as Avicenna, the influential medical writer of a later generation, also wrote on alchemy. He adopted the mercury/sulfur theory of metals, but rejected the possibility of transmutation. Invoking the Aristotelian distinction between the natural and the artificial, he claimed that alchemical processes could never completely replicate the production of gold in nature.

As scholars in the Latin West began the massive task of translating Arabic works into Latin, they included many treatises on alchemy. Knowledge of the works of Arabic alchemy enhanced the existing European craft skills in metal, glass, and dyes. Because the Arabic works had such great authority, many European alchemists produced treatises that they ascribed to a figure they called Geber, leading some readers to believe that Jabir was actually the author of these works. In fact, recent scholarship has established that an obscure Italian Franciscan monk, Paul of Taranto (thirteenth century), was actually the author of some of the most influential treatises attributed to Geber. Paul, or pseudo-Geber, was more interested than the Arab writers in discovering the true causes of chemical phenomena. In the *Summa perfectionis* (*The Sum of Perfection*), sometimes regarded as the bible of the medieval alchemists, he systematically described existing knowledge about metals and minerals as well as the methods for purifying and working with them. He devoted a significant amount of discussion to the transmutation of base metals into gold and the production of the philosophers' stone. He based his theory on a combination of the mercury/ sulfur theory and the Aristotelian theory of *minima*. He used the behavior of the *minima* to explain the observed properties of different kinds of matter as well as chemical changes.

Other medieval thinkers who produced alchemical treatises included Albertus Magnus (1193–1282), Roger Bacon (1214–85), and John of Rupescissa (ca. 1310–ca. 1364). Bacon argued that alchemical gold was better than natural gold and that alchemy could produce a universal medicine capable of curing all diseases. John believed that alchemy could be used to extract the healing "quintessence" from metals dissolved in mineral acids and thus advocated the use of chemical medicines, a suggestion that departed from the traditional emphasis on herbal remedies.

## Describing the World: Natural History

Ptolemy studied the heavens and the earth, but Aristotle focused on the earth's inhabitants. Traditional natural history as a discipline—the study and description of things in the natural world—traces its roots to Aristotle's books *The History of Animals*, *The Parts of Animals*, *The Movement of Animals*, and *The Generation of Animals*. The central concept of an animal's form or soul governed Aristotle's approach to natural history. Accordingly, the full expression of that form realized itself in the mature animal. Aristotle distinguished among groups of animals by describing the essential characteristics of animals rather than by seeking the relationships among different species. Later writers interpreted his descriptions as forming a scheme of classification, although Aristotle himself did not develop such a system.

Aristotle described animals according to their mode of reproduction and according to whether or not they have blood. Animals with blood included hairy quadrupeds, birds, reptiles, amphibians, and fish. Hairy quadrupeds give birth to live young. The other kinds of animals reproduce by means of eggs. He distinguished those with perfect eggs (eggs that have hard shells) from those that have imperfect eggs (eggs that have no shells). Similarly, he divided bloodless animals between those with perfect eggs (octopus, squid, and crustaceans) and those with special kinds of eggs (insects, spiders, and scorpions). He thought that some creatures, like mollusks, are produced from generative slime. He regarded others, such as sponges and jellyfish, as products of spontaneous generation.

Aristotle took a teleological approach to natural history—that is, he believed that the parts of animals, as expressions of their forms, are suited to certain ends. For example, eyelids and eyelashes exist for the protection of the eyes, and the lungs exist to temper the heat of the body. Just as the parts of an animal are expressions of the underlying form, so, in the process of reproduction, the form of the animal passes from the parents to their offspring. The production of offspring that resemble their parents is the end or final cause of this process. According to Aristotle, this finality in the natural world is not the product of intelligent design; rather, the unfolding of nature is an end itself, the result of matter, form, and the four causes as manifested in the world of living things.

Aristotle's student, disciple, and successor as director of the Lyceum, Theophrastus (ca. 380–287 BC), extended his master's project into the plant world. As Aristotle had done for animals, Theophrastus described and categorized many kinds of plants. He based his descriptions on careful observation, although he did not indulge much in theorizing or explaining the phenomena he observed.

Pedanius Dioscorides (ca. AD 40–90), a Greek physician living in Asia Minor, undertook a more specialized study of plants, focusing on medically useful plants. Serving as a surgeon with the Roman army of Emperor Nero, he traveled through much of Europe and North Africa, where he observed hundreds of plants. He wrote a compendium of this information in *De materia medica* (*On Medical Matters*), which served as the authoritative text on herbal remedies throughout the Middle Ages.

Unlike Aristotle, who sought to understand the nature of things, the Roman writer Pliny (Gaius Plinius, 23–79 AD) wanted to describe all of the things in the world. He reportedly traveled with a secretary who wrote down his thoughts and his descriptions of various creatures—real and legendary. His insatiable curiosity led him to the volcano Mount Vesuvius, where he was killed during the famous eruption of 79 AD that destroyed Pompeii. Pliny thought about nature in general terms and believed that some kind of deity governs it. To describe the natural world, he wrote *Natural History*, which runs to ten volumes in a modern edition. Basing his descriptions on the writings of others as well as his own observations, Pliny described animals in detail, revealing Nature's skill and perfection in her creations. He often stressed the relationships between animals and humans, and his accounts often included fantastic as well as carefully observed characteristics. As an example of his approach, consider his description of the beaver, which hunters valued for its testicles, which they believed had special powers.

The beavers of the Euxine, when they are closely pressed by danger, themselves cut off the same part [their testicles], as they know that it is for this that they are pursued. This substance is called castoreum by the physicians. In addition to this, the bite of this animal is terrible; with its teeth it can cut down trees on the banks of rivers, just as though with a knife. If they seize a man by any part of his body, they will never loose their hold until his bones are broken and crackle under their teeth. The tail is like that of a fish; in the other parts of the body they resemble the otter; they are both of them aquatic animals, and both have hair softer than down.[1]

In the Christian Middle Ages, key assumptions about plants and animals developed from a combination of classical sources and references to the Bible. According to the first chapter of Genesis, God created all the living things on earth and in the sea, each "after its kind." All these creatures had "kinds," or essences. Adam's job in the Garden of Eden was to name all the animals. Before the fall from grace into sin, Adam had full access to God's knowledge. Thus he was able to give them names according to their real natures. The names meant something. Because of the emphasis on natures, the biblical account influenced the study of the natural world, reinforcing the Aristotelian tradition, in which the concept of nature or form played a key role. However, Pliny was the source of most medieval accounts of natural history.

Early Christian writers adopted many of Pliny's descriptions and converted them into parables illustrating points of doctrine. The anonymous *Physiologus*, one of the most influential Christianized books of beasts, written in the middle of the second century, contains numerous examples of this practice. Very popular during the Middle Ages, this book was translated from the original Greek first into Latin and then into many other languages. In this work, descriptions of the characteristics and habits of animals metamorphose into Christian allegories. Here is how the author used and transformed Pliny's description of the beaver:

There is an animal called the beaver who is extremely inoffensive and quiet. His genitals are helpful as a medicine and he is found in the king's palace. When the beaver sees the hunter hastening to overtake him on the mountains, he bites off his own genitals and throws them before the hunter. If another hunter happens to pursue him later on, he throws himself on his back and shows himself to the hunter. And the hunter, seeing that the beast has no genitals, departs from him.

O, and you who behave in a manly way. O citizen of God, if you have given

1. Pliny the Elder, *The Natural History*, ed. John Bostock and H. T. Riley (London: H. G. Bohn, 1855–57), pp. 2297–98.

to the hunter the things which are his, he no longer approaches you. If you have had evil inclinations toward sin, greed, adultery, theft, cut them away from you and give them to the devil. The Apostle said, "Pay all of them their dues, taxes to whom taxes are due, honor to whom honor is due," and so on [Rom. 13:7]. Let us first throw the disgraces of sins which are within us before the devil, for they are his works, and let us give to God the things which are God's, prayers and the fruit of our good works.[2]

## Explaining the Human Body: Hippocratic and Galenic Medicine

In addition to the descriptions and explanations of the heavens, the earth, and the earth's inhabitants by astronomers, philosophers, and natural historians, ancient medical thinkers explored a very important class of subjects, human beings and the explanation of the human form. There were three major streams of medical thinking in the Greek world: a tradition of healing associated with the cult of the mythical demigod Asclepius; the tradition of Hippocratic medicine; and a tradition of medicine based on physiology and anatomy, developed by Galen of Pergamon (AD 129–216?). Both Hippocratic and Galenic ideas influenced medicine through the Middle Ages and into early modern times.

According to ancient traditions, Hippocrates (ca. 460–377 BC) wrote some sixty works. In fact, other authors who lived and wrote during the two hundred years following his life were responsible for much of the output ascribed to him. The Hippocratic Corpus, as it is called, covers many aspects of medicine, including the diagnosis and treatment of various ailments as well as medical ethics. The theory of the humors underpinned Hippocratic medicine. This theory posited the existence of four bodily fluids: blood, phlegm, yellow bile, and black bile. The balance or imbalance of these "humors" results in health or illness. Two of the primary qualities form each of the humors. In this way, they resemble the Aristotelian elements. Accordingly, blood, like air, is moist and hot; phlegm, like water, is moist and cold; yellow bile, like fire, is dry and hot; and black bile, like earth, is dry and cold. The Hippocratic physicians related the humors to the elements, the seasons, and the Aristotelian primary qualities.

Imbalances could result from climate, improper nutrition, and other environmental factors. Although the body tends to correct imbalances on its own, diet, exercise, and occasionally drugs can help to restore and maintain balance. Both the theory of the humors and the notion of a non-interventionist therapy influenced medical thinking for centuries. Hippocratic medicine and

2. *Physiologus*, trans. Michael J. Curley (Chicago: University of Chicago Press, 1979), p. 52.

the theory of humors remained popular in the Arab world and persisted in Europe through the Middle Ages and the Renaissance.

Another tradition in Greek medical thinking led to the dissection of human bodies as a way of learning how the body works. The opportunity to perform dissections on human corpses varied, depending on prevailing religious attitudes. Herophilus of Chalcedon (ca. 350–280 BC) and Erasistratus of Chios (ca. 310–250 BC) performed dissections and possibly some vivisections in the Egyptian city of Alexandria, where the Greek colonial rulers had established an extensive library and museum that became important centers of research for some eight centuries. Their studies of human anatomy included descriptions of and early experiments on the cardiovascular system as well as some studies of the brain and nervous system.

In the second century AD, Galen, a successful practitioner, wrote on many aspects of medicine but became especially influential in anatomy and physiology. He described methods of performing anatomies and described the human body in detail. Except during the time he served in Pergamon as physician to the gladiators, Galen himself had limited access to human anatomy, the result of a taboo against touching corpses and cutting into bodies. Therefore, he projected animal anatomy onto the human frame. Although he was aware of the errors that this method might produce, his medieval followers were less meticulous and perpetuated a number of errors. Galen's work resembled Aristotle's in important ways. His physiology was explicitly teleological. He believed that by showing that every part of the body has its use, he had put forth a strong argument against the idea that the universe runs according to chance. Rather, he believed that Nature displays wisdom and intelligence.

Galen based his physiology on the idea that heat is the source of the body's vitality. The vital functions are carried out by three spirits—vegetative, animal, and rational. In the human body, the three principal organs—the liver, the heart, and the brain—modify *pneuma* (an air-like substance) and distribute the resulting spirits through the body by means of three types of vessels: veins, arteries, and nerves. The liver modifies pneuma to form the nutritive soul or natural spirits, which support growth and nutrition. The heart and arteries maintain vital heat, which they distribute through the body along with vital spirits. The brain refines vital spirits into animal spirits, distributed through the nerves to sustain sensation and motion.

According to Galen, the heart occupies a central position between the system of veins and the system of arteries, systems which he thought were completely separate from each other. Earlier anatomists had noted their differences in structure. Galen thought that these differences in structure pointed to differences in function. He thought that the right side of the heart is connected to the venous system and the left side is connected to the arterial system.

The function of the venous system is to provide nutrition to all parts of the body. A person ingests food, which is then processed in the stomach. The portal vein carries the digested food to the liver. The liver refines the digested food, transforming it into blood, which then ebbs and flows through the veins to provide nutrients to all parts of the body. Various organs attract the blood, and that attraction causes the ebbing and flowing. Waste products from the liver's refining process pass into the right atrium of the heart via the vena cava. From the right atrium, they pass into the right ventricle and then to the lungs, via what Galen called the arterial vein (the pulmonary artery). The arterial vein has the structure of an artery, but because it is on the right side of the heart, Galen considered it to be part of the venous system. When the blood reaches the lungs, the waste products are expelled from the body.

The left side of the heart connects to the arterial system via the aorta. Air, processed in the lungs, becomes pneuma, an airy spirit that animates all the parts of the body. The pneuma is carried from the lungs to the left atrium of the heart by the venous artery (the pulmonary vein). Arterial blood receives pneuma from the heart, making it "more golden" and thinner than venous blood. The aorta actively attracts blood, similarly to the way that a bellows sucks in air. This activity by the aorta explains the pulse. Arterial blood passes from the aorta into the rest of the arterial system, carrying this vitalizing spirit and innate heat to all parts of the body. The blood ebbs and flows in the arterial system, and the heart is active in diastole (its expanded state), when it attracts blood.

Because the venous system needs to be animated and the arterial system needs to be nourished, Galen believed that there must be some small amount of blood from each system entering the other. He hypothesized that there exist tiny pores (*anastomoses* in Greek) in the septum—a thick wall separating the right and left ventricles—through which blood can pass from one system to the other. He thought that he had proven that there must be a connection between the venous and arterial systems because severing an artery will cause the veins as well as the arteries to become empty.

The Greek writings continued to influence medical thinking well into the Middle Ages. Avicenna, who wrote on medicine as well as alchemy and philosophy, produced the *Canon of Medicine*, a comprehensive treatise on medical theory and practice which served as an important vehicle for Greek medicine. The *Canon* incorporated the ideas of Hippocrates, Galen, Dioscorides, and others, supplemented by earlier Arabic medical writers. Avicenna's medical writings greatly influenced the Arabic tradition and were translated into Latin in the twelfth century, after which they deeply influenced European thinkers, who called him "the Galen of Islam."

The translation of the Greek and Arabic writings into Latin had an enormous impact on the content of the sciences and natural philosophy as well as on the institutional contexts within which study took place. The works of Aristotle, Ptolemy, and Galen dominated the medieval universities because these works, translated first from Greek into Arabic and later from Arabic into Latin, formed the basis of the medieval curriculum. These works defined nature, the classification of the disciplines, and the course of study. Aristotle's natural books defined the discipline of natural philosophy and excluded mathematics and the mathematical disciplines like astronomy and optics. These disciplines were called mixed mathematics because they combined the method of observation with the method and subject matter of mathematics. Medicine had a faculty to itself and was not considered to be part of natural philosophy, even though it incorporated many natural philosophical ideas. Natural history did not find a place in the university curriculum at all. Neither did alchemy, which its practitioners pursued in other contexts.

Theology was fundamental to natural philosophy in the Middle Ages and beyond, because Aristotelian philosophy was used to explain aspects of Christian doctrine. Natural philosophy studied God's creation. The assumption that God designed the world, created it, and continues to care for it with his providence underpinned most discussions of the natural world. For example, medieval natural philosophers modified Aristotle's claim that the world is eternal in order to allow for its creation by God. They used the Aristotelian theory of matter and change to explain the real presence of Christ in the bread and wine of the Eucharist. The rationality and immortality of the human soul informed discussions of the differences between human beings and animals. Natural philosophy included topics that we would consider theological, not because an oppressive church mandated their inclusion, but because some theological topics were part and parcel of natural philosophy.

After the middle of the fourteenth century, circumstances began to change, and changes in approaches to understanding the world followed. Literary scholars—starting in Italy—sought the original Greek and Latin versions of ancient works, versions that had not undergone the ravages of multiple translations. Doctrinal and political controversies within the Church led to religious restlessness. And the discovery of lands unknown to classical writers contributed to a gradual erosion of traditional ways of understanding the world.

## 2 Winds of Change

*Searching for a New Philosophy of Nature*

During the sixteenth and early seventeenth centuries great changes rocked European intellectual life. Many thinkers actively and explicitly searched for a new philosophy of nature to replace Aristotelianism, which fell into disfavor as the result of the confluence of several major developments. Renaissance humanism, the Protestant Reformation, the exploration of the New World, and Copernican astronomy each in its own way contributed to the erosion of the traditional worldview.

### Restoring the Old to Establish the New: Renaissance Humanism

Starting sometime in the late fourteenth century, mostly outside of the universities, scholars became increasingly aware of the textual corruption of the ancient books that had passed through many translations on their journey from the ancient world to the Latin West. As we have seen, the original Greek texts had been translated first into Arabic in the eighth and ninth centuries and then into Latin in the twelfth and thirteenth. Errors of translation and copying had inevitably crept into these texts. These errors accumulated as the books underwent repeated translations and copying by hand. In many cases Latin translators simply transliterated Arabic terms because no equivalent Latin word existed. Words such as "algebra," "algorithm," "alchemy," "elixir," and "alcohol"—words that remain in our language today—are reminders of these translations.

Troubled by these textual problems, many scholars tried to solve them by searching for original versions of ancient texts, versions that had not suffered the consequences of successive translations. During the first phase of this humanist movement, as it came to be called, scholars rummaged through monasteries and libraries searching for the oldest copies of Greek and Latin texts. Comparing different manuscripts of the same work, they produced what they considered to be the best and most accurate editions of the classical texts.

Because they regarded these ancient works as superior to the productions of their recent past, the humanists sought to emulate the language and style of classical writers. They coined the term "Middle Ages" to signify a period of cultural decline between the Golden Age of Greece and Rome and what they considered to be the rebirth of culture in their own time. The humanists particularly admired the writing of the Roman orator, rhetorician, and philosopher Marcus Tullius Cicero (106–43 BC) and imitated his style of Latin prose, which was far more elaborate than the simpler Latin style that had developed by medieval times. They also believed that the antiquity of a text was the measure of its importance, a spinoff from their conviction that the oldest manuscripts of ancient texts are likely to be the closest to the originals. This attitude influenced their vision of history as well as their detailed scholarship.

The early humanist movement emphasized literary works in contrast to the logic and philosophy that formed the basis of the curriculum of the medieval universities. Later humanists applied their scholarly skills to philosophical and to scientific texts. Many alternatives to Aristotelian philosophy thus became available to natural philosophers, who were actively seeking a new philosophy of nature to replace that of Aristotle. These restored works provided the intellectual matrix within which scholars developed new ways of understanding the world.

Sometimes humanist scholarship produced startling results. Around 1460, while Marsilio Ficino (1433–99) was immersed in editing a new edition of Plato's *Dialogues*, his patron, Cosimo de' Medici, presented him with some Greek manuscripts that a monk had brought back from Macedonia. Cosimo instructed Ficino to drop Plato and turn his attention to these newly discovered documents. What task could possibly be more urgent than translating the writings of Plato, who was regarded as one of the most important of the ancient Greek philosophers? The books that the monk had delivered were reputedly by Hermes Trismegistus ("thrice-powerful Hermes"), whom Greeks and Romans thought was a demigod, the son of a union between a mortal woman and the Egyptian god Thoth—the scribe of the gods and the god of wisdom. The books' references to the Savior and to recognizable doctrines of various Greek philosophers led humanists to regard the Hermetic writings as very ancient—as ancient as the books of Moses—and prophetic.

The truth is far less exciting. These treatises actually dated from the second and third centuries AD, a fact that Isaac Casaubon (1559–1614), a distinguished scholar of Greek, established in 1614. Casaubon revised the date of the Hermetic treatises by closely analyzing the language and references in the texts. He showed that the Hermetic writings contain words that had not appeared in Greek before the Christian era. Despite Casaubon's meticulous scholarship, the treatises continued to have a strong following from the time of Ficino through the seventeenth century.

The Hermetic treatises described a magical view of the universe, one based on an astrological cosmology. They included an account of the creation of the world similar to the one found in Plato's *Timaeus*. Describing a world in which the stars and planets correspond to particular metals, minerals, plants, and parts of the human body, the Hermetic treatises explained how, on a spiritual level, a magus—an individual skilled in Hermetic magic—could escape the fatal influences of the stars by ascending through the heavenly spheres to achieve union with God. On a practical level, some of the Hermetic treatises described how one could draw desired influences into talismans—objects made of special materials and inscribed with symbolic images—in order to use the celestial powers for mundane ends like attracting a lover or ensuring good health. The correspondence between the macrocosm (the heavens) and the microcosm (the earth) provides the key to Hermetic cosmology. Individual objects on the earth—minerals, plants, and gems—contain the signature of the heavenly bodies to which they supposedly correspond. An adept, who understands the correspondence between the macrocosm and the microcosm as well as the symbolic relationships among things in this world, is able to read these signatures and thus to gain an understanding of the specific correspondences at play. Various Hermeticists enriched this tradition by adding elements of Pythagorean numerology and the Jewish Cabala—traditions that ascribed symbolic meaning to numbers and words—to the Hermetic philosophy. This tradition lies at the root of present-day "New Age" beliefs.

The invention of printing in the middle of the fifteenth century ensured that the restoration of the ancient texts would be permanent. Before printing, when all texts were copied by hand, the editorial efforts of the humanists often went for nothing. For example, in 1417, the humanist Poggio Bracciolini (1380–1459) discovered a manuscript of *De rerum natura* (*On the Nature of Things*) by the Roman poet Titus Lucretius Carus (99–55 BC). This work, which proposed an atomic theory of matter, would eventually have a significant influence on seventeenth-century natural philosophy. Poggio's work circulated in manuscript, but only in a few copies, and soon the poem once again fell into obscurity. The survival of Lucretius' poem was not guaranteed until it was printed in 1473.

Eventually scholars used humanist methods to restore the classical writings about various sciences and medicine, and the printed editions of these works stimulated new research. Pliny's *Natural History*, manuscripts of which had circulated widely in the Middle Ages, was printed in 1469. A printed version of Gerard of Cremona's (1114–87) twelfth-century translation of Ptolemy's *Almagest* appeared in 1515. Galen's medical writings in Greek followed in 1525. The works of the Greek mathematician Archimedes (287–212 BC) appeared in the middle of the sixteenth century. A Greek version of Euclid's *Elements* appeared in Basel in 1533, followed by a Latin translation in 1572. The availability of these texts led to renewed interest in these subjects and is at least partly responsible for the flowering of natural philosophy and the sciences in the seventeenth century. The study of medicine, particularly human anatomy, provides a powerful example of how the humanist recovery of ancient texts influenced the subsequent development of the sciences.

Although Galen had written extensively about procedures for dissecting and studying the human body, he had only extremely limited access to human cadavers. Only during the Hellenistic period (fourth through first centuries BC), in Alexandria, had actual dissection of human bodies as a method of research taken place. The medieval medical curriculum actually incorporated the dissection of human bodies in the fourteenth century, but it served to illustrate the ancient anatomical texts rather than to gather new information about the human body.

The illustrations of anatomy in textbooks from the fifteenth and early sixteenth centuries underscore this attitude as well as the distance between the professor, a literary scholar reading an ancient text, and the barber-surgeon, a tradesman who actually dissected the cadaver. Reading from the ancient books, the professor of medicine presided over the event, far above the cadaver, which lay on a table below his podium. A third person, called an *ostensor* or demonstrator, pointed to the relevant part of the body and guided the work of the barber-surgeon who did the actual cutting. Barber-surgeons, as tradesmen, belonged to guilds—associations of artisans or tradesmen. Tradesmen and professionals occupied different social classes. Professors, as professionals, did not work with their hands. They read the works of Galen while tradesmen performed the actual dissection. The resulting division of labor, and the professor's distance from the cadaver, meant that the professor would not become aware of inconsistencies between the anatomical descriptions in Galen's books and the actual structure of the human body.

How can one explain this attitude toward anatomy? In addition to the social factors that separated the scholarly work of professors from the manual labor of tradesmen, medical theory also contributed to these practices. Because the Hippocratic theory of humors dominated medical practice, physicians did not consider knowledge of anatomy relevant to issues of sickness and health.

Although Galen himself had regarded anatomy as a key to a rational understanding of the human body, he did not believe that it served to save human
lives. The combined impact of these two attitudes reinforced the view that
the study of human anatomy had no practical utility. Consequently, pursuing
anatomical research seemed pointless.

The humanist recovery of Galen's writings ultimately revolutionized the
discipline. Andreas Vesalius (1514–64) revived the approach to anatomy Galen
had advocated in his book *On Anatomical Procedures*. Johann Guenther von
Andernach (1487–1574) and Jacobus Sylvius (1478–1555), Vesalius' teachers at
the University of Paris, undertook the recovery of Galen's books. Although
Guenther advocated Galen's methods of dissection, he did not practice what
he preached. Vesalius later wrote of Guenther, "I would not mind having as
many cuts inflicted on me as I have seen him make on either man or other
brute (except at the banquet table)."[1] In contrast, Sylvius actually performed
some of his own dissections and became aware of differences between what he
observed and what he read in Galen's texts. Unwilling to challenge the ancient
authority, Sylvius explained the discrepancies by suggesting that the human
body had actually changed in the 1400 years since Galen had made his observations. Sylvius introduced Vesalius to the practice of dissection. During his
third year in Paris, Vesalius performed the anatomies himself. The difficulty
of obtaining human cadavers reportedly led Vesalius and his students to rob
graves to obtain specimens. They found a good supply of human bones where
the Cemetery of the Innocents (for victims of the plague) was relocated during
the moving of the city walls in Paris.

In 1537 Vesalius assumed the chair of anatomy and surgery at the University of Padua, the most prestigious medical school in Europe. Abandoning
traditional practice, Vesalius performed his own dissections as he lectured. The
new proximity of hand and eye led him to discover errors in Galen's account
of the human body. In stark contrast to Sylvius' conservatism, Vesalius was
willing to acknowledge that Galen had, in fact, made errors.

In order to facilitate the teaching of human anatomy, Vesalius published the
*Tabulae anatomicae* (*Anatomical Pictures*) (1538), six large anatomical drawings
to be used by medical students. He did not intend the drawings to replace dissection and direct observation of the human body but, rather, designed them
to illustrate the readings. This book was the first exposition of the Galenic
physiological system and the first significant attempt to make use of detailed

---

1. As quoted by J. B. deC. M. Saunders and Charles D. O'Malley, *The Illustrations from the
Works of Andreas Vesalius* (Cleveland: World Publishing, 1950), p. 13.

## An Image from Vesalius' *Tabulae anatomicae*

In the *Tabulae anatomicae* Vesalius carefully labeled the parts of the body, using terms from Latin, Greek, Hebrew, and Arabic, as no standardized nomenclature existed. Vesalius drew at least half of the plates himself, and they reveal that he was in the midst of a transition from accepting Galen's authority to relying on the authority of his own observations. The drawing reproduced here makes this tension in his thinking abundantly clear. When he was focusing on the urinary-genital system in a small drawing on the side of the picture, he accurately depicted the liver in the background as having two lobes. On the same sheet of paper, however, he drew a five-lobed liver, for when his focus was the liver, Galenic presuppositions colored his thinking.

■  Reprinted from Andreas Vesalius, *Tabulae anatomicae sex* (Venice: B. Vitalik, 1538), table 1.

drawings in a book on anatomy. Obviously printing, which made the accurate reproduction of identical illustrations a reality, was crucial in making possible the publication of such a work.

Vesalius' work reached its culmination in *De humani fabrica corporis* (*On the Fabric of the Human Body*) (1543), a richly illustrated, complete account of human anatomy based on his own observations. Vesalius described the body in great detail, correcting many of the errors he had found in Galen and in medieval sources. He designed the book as a teaching manual, providing careful descriptions of his methods of dissection so that a reader could repeat what he had done. Vesalius approached anatomy in the same order that Galen had recommended, starting with a description of the skeleton, which formed the framework for the body, then describing the muscles, the vascular system, the nerves, the gastrointestinal system, various membranes and glands, and finally the brain. His actual practice of dissection differed from his instructions. In a time before the invention of formaldehyde or other preservatives, it was necessary to dissect the abdominal cavity and inner organs—the most likely to decompose—before approaching the muscles and bones. Because Vesalius considered structure to be the key to function, he made less use of final causes or purposes than had Galen, whose anatomy was deeply teleological. He also made extensive use of comparative anatomy as a way of highlighting the details of human anatomy.

*De fabrica* is striking for the high quality of its illustrations, which a professional artist produced. Using the illustrations, Vesalius continued to address the problem of nomenclature by standardizing Latin or Greek names for anatomical parts and by labeling them with letters or numbers, which then appeared in the corresponding part of the text. This technique of explicitly linking the illustrations to the text was an important innovation.

In approaching anatomy as he did, Vesalius broke from a centuries-long tradition of commenting on ancient texts by insisting on making his own observations and thereby putting the authorities to the test. These changes in the method and use of dissection, as well as the innovation of new methods of teaching, were possible only after the original Galenic texts became available for scrutiny.

## Renewing the Church: The Protestant Reformation

Prior to the sixteenth century, all Christians in Western Europe belonged to the Roman Catholic Church. The Church had been the primary centralizing institution in the Latin West during the Middle Ages, controlling most of the educational institutions and serving as the theological authority for western

European Christians. Although differences and debates erupted within the Church, no lasting challenge to its authority arose until the early sixteenth century. Similar in some ways to Renaissance humanists, religious reformers in the sixteenth century wanted to improve the state of religion, which they believed had become corrupted during the Middle Ages. To purify Christianity from medieval accretions, they sought a return to the beliefs and practices of the early Church.

Martin Luther (1483–1546), a man who had deep spiritual concerns, found some of the crassly materialistic—even commercial—practices of the Church offensive. In 1517, Luther, an Augustinian monk, posted ninety-five theses on the door of the Castle Church in Wittenberg and wanted to debate these theses with the ecclesiastical authorities. He emphasized the importance of faith and the authority of the Bible rather than outward display. Although his initial protest was a call for reform within the Church, escalating controversy ultimately led to his excommunication from the Roman Church and the establishment of a new, Lutheran church, the theology and practices of which departed significantly from traditional Catholicism.

Basing religion solely on faith and the Bible, Luther advocated what he called a "priesthood of all believers." This slogan expressed his view that any believer is competent to interpret Scripture, a radical departure from the Catholic tradition that vested the authority of interpreting Scripture to the clergy. The medieval Bible was a Latin text that had never been translated into the languages of ordinary people. Luther, who considered faith and Scripture to be the only sources of religious authority, translated the Bible into his native German in order to make it accessible to all believers. He also modified the number and nature of the sacraments, and he permitted priests to marry.

Many of the German princes—in the sixteenth century what is now modern Germany consisted of some 1,700 separate principalities and independent cities—adopted Lutheranism as the official religion of their domains in order to appropriate the vast lands held by the Church. They also wanted to assert secular authority without sharing it with Rome. So the spread of the Reformation had as much to do with politics and economics as it did with religious belief.

Luther attracted many followers, and in the next generation, a host of other reformers emerged. The most significant of these reformers was Jean Calvin (1509–64). Like Luther, Calvin broke from the Roman Church, but he differed from Luther on many points of doctrine and practice. Calvin emphasized knowledge of God, something that he thought could be enhanced by the study of the creation.

Like the humanists, the reformers looked to the past—in this case, the early Church—to find ways to move forward. By the middle of the sixteenth century, the reformers had made such deep inroads into Christendom that the Roman Church convened a council of bishops, the Council of Trent, which met between 1545 and 1563. Its purpose was to determine how the Church would respond to the reformers' challenges. In the end, the council rejected many of the reforms, thus making the schism within the Christian world permanent.

After the Council of Trent, in a movement known as the Counter-Reformation, the Catholic Church set about fighting Protestantism and eliminating some of the more flagrant abuses within its own ranks. The Society of Jesus, founded in 1540 by the Spanish soldier-turned-priest Ignatius Loyola (1491–1556), served as a spearhead of the Counter-Reformation. The Jesuits, as the society's members were called, became a teaching order and established colleges throughout Europe for teaching the sons of the nobility and for promoting what they considered genuine Christianity. They were well educated and produced many important works in theology, philosophy, and natural philosophy.

The significance of the Protestant Reformation for European history cannot be overstated. It had profound consequences for the history of science: it raised questions about the authority to interpret Scripture, about the foundations of knowledge of the world, and about the relationship between biblical claims and natural knowledge. The reformers had challenged the authority of the Church by claiming that each individual has the capacity to read the Bible and to interpret it for him or herself. At the Council of Trent, the Church reaffirmed its traditional position and declared that the authority to interpret Scripture remained in the hands of the Church. Catholic theologians predicted that the reforms would lead to religious anarchy because there would be no criterion for choosing among the multitude of interpretations that were bound to arise from the priesthood of all believers. The Protestants countered by questioning the source of the Church's authority. The recovery of the writings of the classical skeptical philosophers exacerbated the conflict over scriptural interpretation.

In the 1560s, humanist scholars recovered the *Outlines of Pyrrhonism* of Sextus Empiricus (fl. ca. AD 200), a skeptical attack on the ancient philosophers. Sextus presented arguments that challenge not only our ability to know anything, but even our ability to know whether we know. The famous skeptical arguments often provide the starting point for philosophical accounts of knowledge. For example, skeptics ask how we can know the color of an object when it appears to be differently colored in different kinds of light. They note that an oar partly submerged in water appears to be bent; but when it is completely in the air, it appears straight. So how can we determine the real shape of the oar? Food that dogs find tasty appears disgusting to us—so is that food tasty or not? Given such differences, how can we trust our senses to tell us what things are really like? And how can we possibly justify such claims?

Both Catholics and Protestants used these skeptical arguments in their ongoing debates about how to determine a rule of faith and how to establish a criterion for deciding among competing candidates for a rule of faith. Catholics questioned the Protestants' ability to agree on such a rule: differences among individual believers, they argued, would lead inevitably to different ways of interpreting the Bible. Protestants, for their part, questioned the grounds on which the Catholic Church claimed to have sole authority in such matters.

By the beginning of the seventeenth century, skepticism spread beyond discussions about faith and the Bible, leading to a more general skeptical crisis in which all claims to knowledge were subjected to skeptical critique. In particular, philosophers criticized the method of Aristotelian natural philosophy and its claim to achieve certain knowledge about the essences of things. Because natural philosophers no longer considered the traditional Aristotelian method to be a reliable source of knowledge, they tried to develop new approaches to the knowledge of nature, approaches that they thought were immune to skeptical attack.

Questions about biblical interpretation raised other problems as well. What is to be done when a biblical passage contradicts the claims of natural knowledge? Who has the authority to decide? This question, with which theologians had been wrestling practically since the establishment of the Church, became acute during Galileo's conflict with the Church in the seventeenth century. The recovery of Aristotle's writings in the twelfth and thirteenth centuries had directly posed the question to theologians in the universities. For example, Aristotle had argued that the world is eternal, a statement that flatly contradicts the biblical doctrine of creation. Thomas Aquinas and other medieval theologians had simply modified the Aristotelian view and claimed that divine creation is a matter of faith, not knowable by natural reason. Growing confidence in the new methods of the sciences and widespread skeptical criticism of claims to absolute knowledge in the sixteenth and seventeenth centuries led scholars to find new ways of dealing with such questions.

Natural philosophers often used the metaphor of God's two books—the book of God's work and the book of God's word—to formulate the problem. A tradition going back to St. Augustine (354–430) in the fourth century regarded the study of nature as important for understanding passages of Scripture that refer to the natural world. Scholars who accepted this approach regarded the Bible as containing a spiritual message and interpreted passages that deal with the natural world as allegorical and figurative, thereby reserving for natural philosophy the authority for interpreting the created world. Protestant writers frequently reversed the emphasis, focusing on the literal, historical meaning of Scripture, which could thus guide their understanding of the world. These attempts to find positive relationships between theology and religion, on the one hand, and knowledge of nature, on the other, highlight the important point that the relationship between science and religion was not always one of conflict.

## Exploring New Worlds

Ptolemy's *Geography* described the world known by Greeks of his time. His map divided the world into three continents: Europe, Africa, and Asia. His depiction of Europe extended as far north as the east coast of the Baltic Sea. He depicted the Atlantic coast of Africa as a more or less straight line, wandering off towards the southwest. The east coast of Africa extended south beyond the Horn of Africa before it fell off. Although the map includes representations of China and Southeast Asia, their images did not correspond to anything real. He represented the eastern coast of Asia as bending south and connecting with the east coast of Africa, thus portraying the Indian Ocean as an inland sea having no connection with the Atlantic Ocean.

During the Middle Ages, as Islam extended from Spain to the borders of China and from Arabia to Abyssinia (modern-day Ethiopia), Muslim writers acquired a much more detailed knowledge of the Eurasian and African land masses than the Greeks had possessed. Muhammad ibn Musā al-Khwārismī (fl. ca. 830)—also famous for his contributions to the development of algebra—pursued the mathematical aspects of geography, improving Ptolemy's calculations of the shape of the earth and producing better maps of the known world. Arabic expeditions sailed across the Indian Ocean to the Malay Archipelago and Java. Abū al Raihan Bīrūnī (d. ca. 1050) described these travels in his *Book of the Demarkation of the Limits of Areas*. Later works described voyages as well as improved methods for measuring latitude and longitude. A number of these Arabic geographers lived in al-Andalus (Arabic Spain).

Observations of new places and new things called for new approaches to knowledge of the world. Gradually experience replaced authority as the foundation for knowledge. When Christopher Columbus (1446?–1506) sailed to the New World in 1492, when Vasco Da Gama (1460–1524) rounded the Cape of Good Hope and proceeded to India in 1498, and when Ferdinand Magellan (1480?–1521) sailed around the southern tip of South America and circumnavigated the world in 1521, Europeans became aware of huge oceans and masses of land that Ptolemy had not even mentioned in his *Geography*—despite his reasonably accurate estimate for the size of the earth.

These explorations of the New World, Africa, and Asia had a direct impact on intellectual life, contributing to the breakdown of traditional ideas and traditional authorities. Scholars in both the Middle Ages and the Renaissance had consulted books by ancient Greek and Roman writers as authoritative accounts of geography and natural history. The exploration of lands that the ancients had never seen, the discovery of animals and plants that the ancients had not described, and the sight of human races and cultures previously unknown led scholars to question the accuracy and authority of the classical texts. Perhaps ancient writers did not have the last word on the extent and contents of the world.

At first, explorers saw their discoveries through the lenses of the ancient texts. Columbus thought he had come to China, as there had been no inkling of either North or South America in Ptolemy's *Geography*. Later explorations clearly challenged the accuracy of Ptolemy's maps and firmly established the existence of a huge landmass in the Western Hemisphere. These discoveries undermined the authority of the ancient texts, substituting direct experience as the source of geographical knowledge.

The inadequacy of Ptolemy's maps created an urgent need for better maps and better methods of navigation. In addition to careful observations, mapmakers used mathematical methods to render the new discoveries in visual form. Craftsmen who manufactured the instruments needed to produce maps and to make the observations necessary for navigation played a crucial role in facilitating the explorations. Consequently, practical mathematics rose in importance, not only for mapmaking and navigation, but also for surveying the newly discovered territories.

The discovery of the New World challenged the ancient accounts of natural history as well. The early explorers—Christopher Columbus and Amerigo Vespucci (1451–1512)—described the animals they encountered in the New World in terms familiar from classical authors like Pliny and the more recent account that Marco Polo (1254–1324) had written about his overland travels to China. Credulous about reports of animals such as the "monkey-cat" of Peru (supposedly the result of an "adulterous union" between a monkey and a cat), European naturalists encountering previously unknown species for the first time tended to compare them with creatures described by the classical authors.

The monkey-cat seemed no less weird than the griffin, a mythological creature possessing the head and wings of an eagle and the body of a lion. Gradually, as explorers, soldiers, and missionaries began to trust their own observations more than ancient texts, they realized that many creatures inhabiting the New World had been unknown to the classical writers and were different from familiar animals or plants inhabiting the Old World. The explorations contributed to a revision of the attitude that had granted authority to classical antiquity, replacing it with a new respect for experience and observation. Not only were Renaissance scholars willing to supplement the classical accounts, but they also became more critical of reports from travelers and others.

## Reforming the Heavens: Copernican Astronomy

Ptolemaic astronomy and Aristotelian cosmology remained dominant through the European Middle Ages, while at the same time Arabic astronomers were making important modifications to these theories. Early modern European astronomers were well aware of the incompatibility between the Aristotelian and Ptolemaic systems, an incompatibility that the Arabic astronomers had demonstrated. An ongoing debate about this problem provided the context for further developments in astronomy. Averroes' criticism of Ptolemaic astronomy precipitated this debate in Europe. Like other Arabic astronomers, Averroes and his followers thought that astronomy should provide a real description of the heavens. Because Averroes accepted the principles of Aristotelian cosmology, he believed that the movements of the heavenly bodies should be explained by mechanisms using only uniform circular motion. He criticized Ptolemaic astronomy, despite its ability to make accurate predictions, because it faced possibly insurmountable problems in determining which of several equivalent mathematical constructions is true. Besides, neither epicycles nor eccentrics (and especially not the equant) were consistent with the requirement of uniform circular motion. Consequently, Averroes came to the radical conclusion that astronomy in the Ptolemaic mode is illegitimate and should not be pursued. European astronomers knew about the debates surrounding Averroes' critique of Ptolemaic astronomy. Dealing with these issues led to important, even radical changes in astronomy.

The recovery of Ptolemy's texts and their translation from Greek into Latin in the middle of the fifteenth century stimulated further consideration of these issues. In the last half of the fifteenth century, some European astronomers, following in the tradition of Ptolemy's *Planetary Hypotheses*, which they knew from Arabic sources, attempted to give physical explanations of the planetary motions. Georg Peurbach (1423–61) popularized this approach in his *Theoricae novae planetarum* (*New Theories of the Planets*), which he completed by 1454 but which did not appear in print until 1472. He used three-dimensional orbs (solid, spherical shells) to produce the motions of the planets and then used Ptolemaic calculations in the manner of the *Almagest* to predict planetary positions. This book became the most widely read introduction to astronomy in the sixteenth century. In collaboration with Johannes Regiomontanus (1436–76), Peurbach wrote the mathematically rigorous *Epitome of the Almagest*.

These works provided the technical background for the revolutionary system Nicholas Copernicus (1473–1543) developed. Copernicus was born in East Prussia (now part of Poland) and attended the University of Cracow, after which he spent eight years in Italy, where he received a degree in canon (Church) law. His studies included mathematics, astronomy, astrology, canon and civil law, and medicine. During his time in Italy, Copernicus learned about the Averroistic attack on Ptolemaic astronomy and, more generally, the Arabic contributions to astronomy. On his return to East Prussia, he assumed administrative duties as a canon in the cathedral at Frauenberg, a position he held for the rest of his life.

Copernicus altered astronomy in ways that precipitated revolutionary changes in mathematical astronomy and physics. He developed a detailed heliocentric (sun-centered) account of the motions of the planets, and he insisted that astronomy describe the physical reality of the heavens. He wrote a summary of his system, the *Commentariolus* (*The Little Commentary*) and circulated it in manuscript before 1514. In this short work, Copernicus outlined his new heliocentric system of astronomy and claimed that it describes physical reality. Like the Arabic astronomers, he insisted on using only uniform circular motions, and he deployed the Tūsī-couple to avoid using the equant or other constructions that violated the principle of uniform circular motion. Although we do not know precisely how Copernicus learned about the details of Arabic astronomy, he clearly knew about them and incorporated them into his work.

By this time, Copernicus had already achieved a reputation as an important astronomer. He was so well-known that a papal commission asked him for advice on calendar reform. The calendar needed to be overhauled because errors in the length of the year had multiplied during the many centuries since the adoption of the Julian calendar in the first century BC, with the result that the calendar was no longer synchronized with astronomical phenomena such as the vernal equinox. Apart from practical considerations, the Church was concerned with the calendar because of the need to determine the proper dates for religious feasts, especially Easter. Copernicus declined to advise the commission, stating that both observational and theoretical astronomy needed to be improved before it would be possible to correct the problems with the calendar. Publishing nothing, he continued working out the technical details of his system until 1539, when Georg Joachim Rheticus (1514–74), a Lutheran professor of mathematics from Wittenberg, came to Frauenberg in order to learn more about Copernicus' new astronomical system. Excited by the theory, Rheticus published a short account of it, the *Narratio prima* (*The First Telling*), in 1540.

During his extended stay, Rheticus encouraged Copernicus to publish the full version of his astronomy, *De revolutionibus orbium coelestium* (*On the Revolutions of the Heavenly Spheres*). Because the book contained numerous geometrical diagrams and tables of observations, the process of printing required constant supervision. In 1642, Rheticus took the manuscript to a printer in Nuremburg and then departed for a teaching position at the University of Leipzig, leaving the supervision of the printing of *De revolutionibus* in the hands of a Lutheran minister, Andreas Osiander (1498–1552). Here the plot thickens.

Osiander inserted an unsigned preface into the book, stating that the major astronomical claims asserted in the book should be taken simply as mathematical hypotheses designed to save the phenomena—that is, to construct mathematical models that account for all the observations and generate true predictions—but not as descriptions of physical reality. Although this preface flatly contradicted Copernicus' statements throughout the book, many readers took it at face value, and very few astronomers accepted the physical reality of heliocentrism until Johannes Kepler (1571–1630), who had read the correspondence between Rheticus and Osiander, revealed the actual authorship of the preface. Copernicus died in 1543 just as the first copies of *De revolutionibus* came off the press.

With the benefit of hindsight we can see that the book precipitated profound changes in astronomy, physics, and cosmology, but Copernicus' goal was not revolutionary. In many ways his approach was, in fact, conservative. For example, he considered the observations of the ancient astronomers to be totally reliable. "We must follow in their footsteps and hold fast to their observations, bequeathed to us like an inheritance. And if anyone on the contrary thinks that the ancients are untrustworthy in this regard, surely the gates of this art are closed to him."[2]

2. Nicholas Copernicus, "Letter against Werner" (1524), in Edward Rosen, *Three Copernican Treatises* (New York: Dover, 1959), p. 99.

In fact, Copernicus did not make many new observations. Indeed, the precision instruments needed for making significantly improved observations did not yet exist. In addition to trusting the ancient observations, he shared the ancient goal of accounting for the motions of the heavenly bodies by using combinations of uniform circular motions. He criticized Ptolemy's system because, in attempting to account for some of the inequalities of planetary motions, Ptolemy had employed the equant, which Copernicus, like many of the Arabic astronomers, considered to be a violation of uniform circular motion. He explained this point in the *Commentariolus*.

> Having become aware of these defects, I often considered whether there could perhaps be found a more reasonable arrangement of circles, from which every apparent inequality would be derived and in which everything would move uniformly about its proper center, as the rule of absolute motion requires. After I had addressed myself to this very difficult and almost insoluble problem, the suggestion at length came to me how it could be solved with fewer and much simpler constructions than were formerly used, if some assumptions . . . were granted me.[3]

Copernicus' new assumptions literally turned the cosmos inside out. He removed the earth from the center of the universe and set it into motion around the sun, which he placed at the center. He gave the earth a second motion, a daily rotation on its axis, in order to explain the daily risings and settings of the sun, moon, planets, and stars. And he showed how the motion of the earth could account for the observed stations and retrogradations of planetary motions. Copernicus thought that this new theory would preserve uniform circular motion more harmoniously and more effectively than Ptolemy's.

Copernicus opened *De revolutionibus* by describing the physical structure of the cosmos. Agreeing with the ancient astronomers, he argued that both the cosmos and the earth are spherical. The heavenly bodies move either in uniform, eternal, and circular motions or in motions compounded of circular motions. He included the earth among these heavenly bodies, disagreeing with most of the ancient astronomers, who had held the earth to be at rest at the center of the cosmos. Invoking what would later be called the principle of optical relativity, he noted that if two bodies are moving relative to each other and the observer is located on one of these bodies, by visual evidence alone it is impossible to determine which body is actually moving. It follows that the earth's daily rotation on its axis can explain the observed daily rising and set-

---

3. Nicholas Copernicus, *The Commentariolus*, in Rosen, *Three Copernican Treatises*, pp. 57–58.

## Copernicus' Depiction of the Cosmos

In Copernicus' revolutionary theory the sun is stationary at the center of the cosmos. The planets and stars all revolve in circular orbits around the sun. The earth is the third planet out from the sun, about which it revolves, now having the same status as the other planets.

■ Reprinted by permission of the publisher from *The Copernican Revolution: Planetary Astronomy in the Development of Western Thought* by Thomas S. Kuhn, p. 162, Cambridge, Mass.: Harvard University Press, Copyright © 1957 by the President and Fellows of Harvard College, Copyright © renewed 1985 by Thomas S. Kuhn.

ting of all the heavenly bodies just as effectively as the daily rotation of all the heavenly bodies around the earth.

In addition to its daily rotation, Copernicus claimed that the earth revolves around the sun, completing one revolution every year. This motion provides an easy explanation for the stations and retrogradations of the other planets. Copernicus regarded his explanation as superior to Ptolemy's. Unlike Ptolemy, who constructed a separate geometrical model for each planet in order to explain these phenomena, Copernicus explained them as natural consequences of his theory without needing to make any additional assumptions.

Displacing the earth from the center of the universe explained other phe-

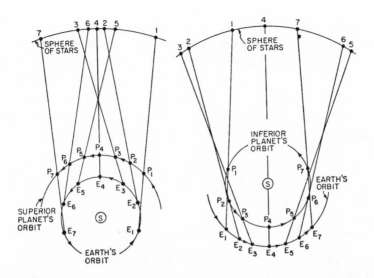

## Stations and Retrogradations in the Copernican System

All planets travel in their orbits from west to east at different rates of speed. From time to time they appear to stop, reverse their motion, travel from east to west, stop again, and resume their motion from west to east. These motions are called stations and retrogradations. Copernicus was able to explain them as a natural consequence of his system.

The diagram illustrates Copernicus' explanation of the apparent motion of the other planets relative to the earth. $E_1$ through $E_7$ are the positions of the earth as it revolves around the sun, and $P_1$ through $P_7$ are the positions of one of the planets beyond earth's orbit at these same moments. Projected onto the sphere of the fixed stars, the planet seems to move in forward motion from 1 to 3, then it seems to reverse its motion from 3 to 4 to 5, at which point it seems to resume its forward motion. An observer sees this backtracking when the earth passes the planet as they each move around the sun at different rates. The observed motion is much like what you see when you pass a car on the highway and it seems to recede behind your car. A similar effect occurs with regard to the planets interior to the earth's orbit (Mercury and Venus).

■ Reprinted by permission of the publisher from *The Copernican Revolution: Planetary Astronomy in the Development of Western Thought* by Thomas S. Kuhn, p. 163, Cambridge, Mass.: Harvard University Press, Copyright © 1957 by the President and Fellows of Harvard College, Copyright © renewed 1985 by Thomas S. Kuhn.

nomena that Ptolemy had dealt with on an ad hoc basis. In Ptolemy's system, because the geometrical construction for each planet is independent of all the others, no criterion exists for determining the order of the planets. For Copernicus, the order of the planets can be determined simply by placing them in the order of the time it takes each planet to complete an orbit of the sun, a number that is known as the planet's period:

**Periods of the Planets**

| | |
|---|---|
| Mercury | 88 days |
| Venus | 225 days |
| Earth | 365.25 days |
| Mars | 687 days |
| Jupiter | 12 years |
| Saturn | 29 years |

The Copernican system easily explained why Mercury is never observed to be further than 29 degrees from the sun and Venus is never observed to be further than 47 degrees from the sun. For Copernicus, these facts need no additional assumptions because in his system Mercury and Venus are interior to the earth's orbit. Consequently, viewed from the earth, they are never more than a limited angular distance from the sun.

Copernicus ascribed a third motion to the earth to account for the seasonal changes in the height of the sun in the sky. Still thinking of the heavenly bodies (including the earth) as attached to spheres, he needed this third motion to keep the earth's axis tilted at a constant angle of 23½ degrees. The third motion keeps the earth's axis parallel to itself as it revolves around the sun. Once astronomers abandoned the theory that the earth and the other planets are embedded in solid spheres but move in empty space, this third motion was not necessary, for the earth's axis would simply maintain its angular position.

Qualitatively the Copernican system seemed simpler and more elegant than any of its predecessors. Copernicus apparently needed only seven circles to solve the same problems for which Ptolemy needed twelve. This appearance of greater simplicity was illusory, however, for when Copernicus worked out the detailed computations of the motions of the planets, his system became just as complex as Ptolemy's. He even needed to use a construction similar to the equant to account for some of the more complex motions of the planets. Since he relied on the same data that Ptolemy had used, his system was no more accurate than Ptolemy's. Greater accuracy would not be possible until extensive, new observations became available.

Although Copernicus' heliocentric scheme seemed to have many advan-

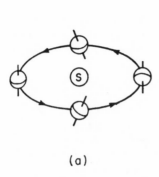

(a)

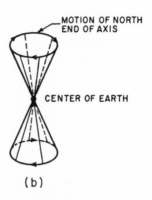

MOTION OF NORTH
END OF AXIS

CENTER OF EARTH

(b)

## Copernicus' "Second" and "Third" Motions

In the Copernican scheme the first motion is that of a planet, the earth, rotating around the sun. The second motion is the daily motion of the earth on its axis. Because Copernicus was still thinking of the earth as attached to a solid orb rotating around the sun, the earth's axis would rotate in a circle over the course of a year, as shown in (a). Copernicus claimed that the earth also had a third motion. The second motion does not keep the earth's axis parallel to itself, so that the conical third motion shown in (b) is required to bring

the axis back into line. With the elimination of the orbs and the theory that the earth revolves about the sun in empty space, this third motion was no longer needed, as the earth maintains its orientation while it revolves around the sun.

■ Reprinted by permission of the publisher from *The Copernican Revolution: Planetary Astronomy in the Development of Western Thought* by Thomas S. Kuhn, p. 165, Cambridge, Mass.: Harvard University Press, Copyright © 1957 by the President and Fellows of Harvard College, Copyright © renewed 1985 by Thomas S. Kuhn.

tages over Ptolemy's, its fundamental assumption of the earth's motion encountered serious objections from traditional physicists and astronomers. Ancient Greek astronomers had already articulated these objections against Aristarchus of Samos (fl. 310–230 BC), who had also proposed a heliocentric theory. In the Middle Ages, similar objections arose in the context of discussions about God's power to create a universe in which the earth moves. In the opening sections of *De revolutionibus*, Copernicus addressed these objections directly. In response to the objection that earth's motion would cause it to fly apart, a phenomenon that is not observed to happen, Copernicus replied that since the heavens are so much larger than the earth, this argument would apply even more strongly

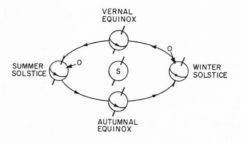

## The Earth's Annual Motion around the Sun

sees the sun more nearly overhead at the winter solstice.

Once astronomers eliminated the solid orbs, they pictured the earth as maintaining its orientation as it revolves around the sun. At all times the earth's axis stays parallel to itself or to the stationary line drawn through the sun. As a result an observer at $O$ at noon in middle-northern latitudes finds the sun much more nearly overhead at the summer than at the winter solstice. An observer in the southern hemisphere

■ Reprinted by permission of the publisher from *The Copernican Revolution: Planetary Astronomy in the Development of Western Thought* by Thomas S. Kuhn, p. 166, Cambridge, Mass.: Harvard University Press, Copyright © 1957 by the President and Fellows of Harvard College, Copyright © renewed 1985 by Thomas S. Kuhn.

to the motion of the heavens. He invoked the principle of optical relativity, noting that by astronomical observations alone we cannot settle the question of the earth's motion. This argument had great significance, for it implied that the question could be resolved only by introducing physical considerations, a procedure that called for crossing traditional disciplinary boundaries and thus revising the Aristotelian classification of the sciences.

Other traditional arguments against the motion of the earth also rested on Aristotelian physics. The earth's motion would cause unattached objects such as clouds and animals to fly off its surface; and bodies in free fall, such as rocks dropped from a tower, would fall away to the west, as the surface of the earth rotates beneath them. Copernicus refuted these objections by asserting that bodies close to the earth's surface share the earth's motion, and so appearances would remain the same, whether or not the earth moves. Copernicus did not explain how these bodies share the earth's motion, but clearly Aristotelian physics could no longer provide an adequate explanation. Whether or not

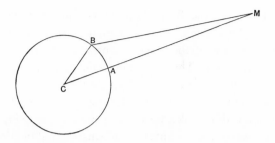

## Annual Stellar Parallax

Because the line between a terrestrial observer and a fixed star does not stay quite parallel to itself as the earth moves in its orbit, the star's apparent position on the stellar sphere should shift by an angle *BMA* as the earth moves between *A* and *B* on its orbit. The maximum parallax occurs at points that are six months apart. Copernican astronomy predicted parallax, but none was observed until the nineteenth century because parallax is not visible with naked-eye instruments.

Copernicus understood the radical nature of his views, his system, if taken as physically real, demanded the creation of a new physics.

The one observable prediction from the earth's moving in an orbit was annual stellar parallax, but no one observed parallax during Copernicus' lifetime. Annual stellar parallax, which is on the order of a third of a second of arc, was not observed until 1838, when Friedrich Wilhelm Bessel (1784–1846) was able to use a newly invented, very powerful telescope. Copernicus did not regard the absence of observed parallax as a refutation of his theory. Instead, he proposed that the stars are much farther from the earth than previously thought, so far, in fact, that the angle of parallax is too small to be observable by the naked-eye instruments available at the time.

Few astronomers adopted Copernican astronomy during the first fifty years following the publication of *De revolutionibus*. Although the close relationship between geocentric Aristotelian cosmology and Christian theology suggests that Copernicus' views should have provoked a hostile response from religious authorities, in fact there was none. Copernicus himself served as a canon in the Roman Catholic Church for his entire life. He dedicated his magnum opus to Pope Paul III, and the book contained an introductory letter from the Dominican priest Nicholas Schönberg, cardinal of Capua, who had learned of Copernicus' new astronomy and cosmology from humanist churchmen acquainted with his work. The ecclesiastical authorities had far more serious concerns at the time. The pope and other high officials in the Church were organizing the Council of Trent, which met from 1545 to 1563, the main purpose of which was to determine how to react to the religious reforms that Martin Luther had initiated earlier in the century. The revolutionary theories of a Polish astronomer in a remote corner of Europe were hardly of great concern.

Although Copernicus clearly believed in the physical reality of his astronomical system and thus broke down the traditional disciplinary boundary between astronomy (a branch of mixed mathematics) and physics (or natural philosophy), Osiander's preface reinforced the traditional division. According to that tradition, astronomers should calculate the observed positions of the heavenly bodies without regard to physical truth, and natural philosophers should search for the real causes of phenomena. This distinction loomed large in the initial reception of Copernicus' work. Several young astronomers at the University of Wittenberg, under the tutelage of Philipp Melancthon (1497–1560), were willing to use Copernicus' methods for calculating planetary positions, preferring his methods to Ptolemy's, since Copernican methods enabled them to avoid using the equant. But they were not willing to embrace his heliocentric cosmology. An important member of this group, Erasmus Reinhold (1511–53) published a set of astronomical tables, the *Tabulae Prutenicae* (*Prussian Tables*) (1551), calculated using Copernican methods. These tables were more extensive, more accurate, and easier to use than the previous Alphonsine tables, which had been compiled in the thirteenth century.

A further breakthrough in astronomy required new and more accurate observations. Tycho Brahe (1546–1601), a Danish nobleman, took on this task as his life's work. At sixteen, he made some observations that revealed the weaknesses of existing data. Observing a rare conjunction of Saturn and Jupiter (a sighting of the two planets very close to each other), he determined that predictions based on the Alphonsine tables failed by a full month and that even the new *Tabulae Prutenicae* missed the mark by two days. This early experience made him realize that a reform of astronomy must be based on more accurate observations. Several other observations reinforced his commitment to this calling. In 1572, a new star appeared in the constellation Cassiopeia. This bright new star shone even during the day. It remained visible for eighteen months and then disappeared. It exhibited no observable parallax, an indication that it was in the region of the fixed stars, part of the heavenly region, in which, according to Aristotle, no qualitative change and no coming-into-being or passing-away was possible. Tycho's observation of this new star (the first of many) chipped away at the foundation of Aristotelian cosmology.

## Tycho Brahe's System

In Brahe's system, the earth is stationary at the center of the universe. The moon and the sun travel around the earth. All the other planets travel around the sun. X is the comet of 1577.

■ Reprinted from Arthur Berry, *A Short History of Astronomy from Earliest Times through the Nineteenth Century* (London: John Murray, 1898), p. 137.

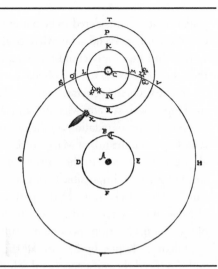

A brilliant comet appeared in the heavens in 1577. Parallax measurements established that this comet moved through the region of the planets, a region in which—according to traditional theory—only uniform circular motion on a solid sphere existed. Tycho realized that this comet would be passing through those solid spheres, a physical impossibility. Eventually he concluded that no spheres exist, thus raising the urgent question of what holds the planets in their orbits. This question again drew a definite connection between astronomy and physics.

Tycho's determination to improve astronomy went beyond these isolated, but very significant observations. With patronage from the King of Denmark, Tycho established an observatory on the island of Hven in the sound between Denmark and Sweden. Calling the observatory Uraniborg (Heavenly Castle), he built an alchemical laboratory, installed astronomical instruments, and provided accommodations for several observers. The huge size and stability of the instruments at Uraniborg enabled Tycho's team of astronomers to make dramatically improved naked-eyed observations. Observing systematically for twenty years, they compiled tables that vastly improved all previous observations. Tycho's observations of stars achieved an accuracy of about one minute of arc. Previous observations had been accurate at best to four minutes of arc. This fourfold increase in accuracy led to substantial improvement in astronomical theory.

In addition to his observations, Tycho also formulated an astronomical

system that he considered to be better than both the Ptolemaic and Copernican alternatives. Tycho acknowledged the many advantages of the Copernican
system, but he simply could not accept the motion of the earth. After much
thought, he constructed a geoheliocentric system.

Tycho's scheme generated observable consequences identical with those of
the Copernican system. Mercury and Venus remain close to the sun, and the
stations and retrogradations of the planets are natural consequences of this system. However, the orb of Mars crosses the orb of the sun, a major change from
all previous systems. In the Aristotelian scheme, such an intersection of orbs
would have been impossible because the orbs were embedded in solid spheres.
But Tycho's observations of the path of the comet of 1577 and the problem
of Mars in his system had led him to abandon the solid spheres. Once again,
physical considerations became relevant to astronomical theory.

Johannes Kepler broke with all the traditional astronomical assumptions
and produced the most successful solution to the problem of the planets that
anyone had proposed since the beginnings of mathematical astronomy. He
was able to accomplish this end by abandoning the assumption that motions
of the heavenly bodies can be described by combinations of uniform circular
motions.

As a young man, Kepler studied for the Lutheran ministry. Because he held
certain unorthodox beliefs, positions both in the ministry and at the university
were unavailable to him, for ordination was required for all university professors. Consequently, he continually searched for employment and patronage.
Kepler's teacher at the University of Tübingen from 1589 to 1594, Michael
Maestlin (1550–1631), introduced him to Copernicanism, even though Maestlin himself did not consistently endorse heliocentric astronomy. Kepler, who
did accept Copernicanism, believed that the study of nature provides knowledge of God's providential plan for the universe. Kepler's belief that God is a
divine geometer who embedded precise mathematical relationships into the
world motivated his emphasis on empirical accuracy: astronomical theory
must describe the heavens with extreme precision.

Kepler's advanced studies in astronomy with Maestlin enabled him to find
a position teaching mathematics in a Lutheran school in the Austrian city of
Graz. As Kepler related the story, in the middle of teaching a class, asking his
students why there are just six planets (Mercury, Venus, Earth, Mars, Jupiter,
and Saturn), no more and no fewer, Kepler realized that the Copernican orbs of
these planets could be inscribed in and circumscribed around the five Platonic
solids (the tetrahedron, cube, octahedron, dodecahedron, and icosahedron).
These solids have the characteristic that each face of each solid is equilateral

and congruent with each of its other faces. Mathematicians had long since demonstrated that five and only five such solids exist. It follows that there are only six spheres that can be circumscribed around and inscribed within them. For Kepler, who held a Pythagorean approach to mathematics—an approach that regards mathematical harmony as causally efficacious—this mathematical fact explained the physical fact that there are precisely six planets.

This model explained the order of the planets, linking the proportions of the orbits to musical harmonies and astrology. This insight resulted in his first book, the *Mysterium cosmographicum* (*The Cosmographical Mystery*) (1596), the first published endorsement of Copernican cosmology by a professional astronomer since the appearance of Rheticus' *Narratio prima* in 1540.

Proud of his book, Kepler sent copies to many European astronomers, including Maestlin, Galileo, and Tycho. Galileo, horrified by Kepler's Pythagoreanism, never replied, thus breaking off their correspondence. Tycho, who in 1599 became imperial astronomer in the court of Rudolph II, the Holy Roman emperor, recognized Kepler's mathematical and astronomical talents and invited him to Prague to work with the data that he had systematically collected for some twenty years at Hven. Kepler seized the opportunity. He joined Tycho in Prague in 1600 and set to work on the planet Mars, the planet whose orbit has the greatest eccentricity (i.e., it deviates from true circularity more than that of any other planet).

Kepler worked on the problem of Mars until 1609, when he published the *Astronomia nova* (*A New Astronomy*), which contains the first two of his three laws of planetary motion. With these laws, Kepler abandoned the ancient assumption that the planetary motions must be explained in terms of uniform circular motion. Moreover, setting the planets in elliptical orbits, not around the center of the ellipse, but about one of the foci where the sun is situated, added urgency to the question of what holds the planets in their orbits. The full title of the *Astronomia nova*—in English, *New Astronomy, Based upon Causes, or Celestial Physics, Treated by Means of Commentaries on the Motions of the Star Mars from the Observations of Tycho Brahe, Founded on the Observations of the Noble Tycho Brahe*—announced Kepler's pursuit of a physical explanation for the motions of the planets. Kepler not only solved the problem of the planets by abandoning the ancient assumptions of uniform circular motion, but he also explicitly revolutionized the relationship between the disciplines of astronomy and physics by calling his new astronomy a "celestial physics."

Examination of Kepler's methods reveals his insistence that his theory account for the observations with complete accuracy. For example, consider his first approach to the problem of Mars. He took four of Tycho's observations

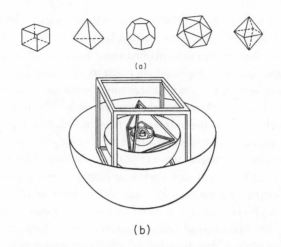

(a)

(b)

## Kepler's Model of the Planetary Orbs

Diagram (a) shows the five Platonic solids in which Kepler believed the Copernican orbs of the planets could be inscribed. From left to right they are cube, tetrahedron, dodecahedron, icosahedron, and octahedron. Their order is the one that Kepler developed to account for the sizes of the planetary spheres. Diagram (b) shows the solids in this application. Saturn's sphere is circumscribed about the

cube and Jupiter's sphere is inscribed in it. The tetrahedron is inscribed in Jupiter's sphere, and so on.

■ Reprinted by permission of the publisher from *The Copernican Revolution: Planetary Astronomy in the Development of Western Thought* by Thomas S. Kuhn, p. 218, Cambridge, Mass.: Harvard University Press, Copyright © 1957 by the President and Fellows of Harvard College, Copyright © renewed 1985 by Thomas S. Kuhn.

of Mars at aphelion and perihelion—when it was furthest and closest to the sun—and hypothesized the size of Mars' eccentricity. After incredibly long calculations, he was able to account for all of Tycho's observations of Mars at aphelion to within two minutes of arc, the margin of error of his data. However, when he tested the theory against Tycho's observations of Mars when it was at quadrature (90 degrees from aphelion), he discovered that his model predicted positions that differed from the observations by eight minutes of arc. He remarked that Ptolemy or Copernicus could have overlooked a discrepancy of eight minutes, because their observations were accurate only within a margin of error of ten minutes.

Since the divine benevolence has vouchsafed us Tycho Brahe, a most diligent observer, from whose observations the eight-minute error in this Ptolemaic computation is shown, it is fitting that we with thankful mind both acknowledge and honor this benefit of God. . . . In what follows I shall myself, to the best of my ability lead the way for others on this road. For if I had thought that I could ignore eight minutes of longitude, in bisecting the eccentricity I already would have enough of a correction in the hypothesis found in ch. 16. Now, because they could not have been ignored, these eight minutes alone will have led the way to the reformation of all of astronomy, and have constituted the material for a great part of the present work.[4]

This error of eight minutes led Kepler to discard calculations that had consumed two years of his time. His insistence on a precise fit between his theory and the data derived from his conviction that a proper astronomical theory describes physical reality. God, the geometer, created a world that embodies precise mathematical relationships.

Kepler's path to his first two laws was circuitous. The error of eight minutes ultimately led him to reject both a circular orbit for Mars and uniform speed around its orbit. He realized that the speed with which the planet travels around its orbit is not uniform. In attempting to determine the rate at which the planet moves, he discovered what became known as his second law, namely, that the planet's radius sweeps out equal areas in equal times.

But he still had to determine the shape of the planet's orbit. Having rejected the circle, he considered the possibility of some kind of an oval. Examining the hypothesis of an egg-shaped orbit, he found it impossible to calculate its area, and exclaimed, "If only it were a perfect ellipse, all the answers could be found in Archimedes' and Apollonius' work."[5] After many fruitless calculations with the oval, Kepler stumbled upon the ellipse, which fit the observations and made calculations possible. His account of these discoveries led him to declare, "The roads that lead man to knowledge are as wondrous as that knowledge itself."[6]

As the full title of the *Astronomia nova* makes perfectly clear, Kepler was committed to finding a physical explanation for the motions of the planets. Still thinking in terms of the Aristotelian assumption that all motion requires

4. Johannes Kepler, *New Astronomy*, trans. William H. Donahue (Cambridge: Cambridge University Press, 1992), p. 286.
5. Kepler to D. Fabricius, 18 December 1604, quoted in Arthur Koestler, *The Sleepwalkers: A History of Man's Changing Vision of the Universe* (1959; London: Penguin, 1964), p. 335.
6. Quoted by Koestler, *The Sleepwalkers*, p. 337.

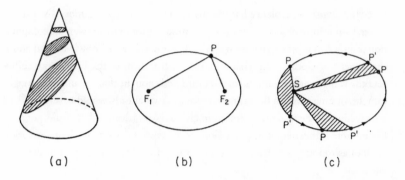

(a)                    (b)                    (c)

## Kepler's First Two Laws

Diagrams (a) and (b) define the ellipse, the geometric curve in which all planets that obey Kepler's first law must move. In (a) the ellipse is shown as the closed curve in which a plane intersects a circular cone. When the plane is perpendicular to the axis of the cone, the intersection is a circle, a special case of the ellipse. As the plane is tilted, the curve of intersection is elongated into more typically elliptical patterns.

A more modern and somewhat more useful definition of the ellipse is given in diagram (b). If two ends of a slack string are attached to two points $F_1$ and $F_2$ in a plane, and if a pencil $P$ is inserted into the slack and then moved so that it just keeps the string taut at all times, the point of the pencil will generate an ellipse. Changing the length of the string or moving $F_1$ and $F_2$ together or apart alters the shape of the ellipse in the same way as a change in the tilt of the plane in diagram (a). Most planetary orbits are very nearly circular, and the foci of the corresponding ellipses are therefore quite close together.

Diagram (c) illustrates Kepler's second law, which governs orbital speed. The sun, $S$, is at one focus of the ellipse, as required by the first law, and its center is joined by straight lines to a number of planetary positions, $P$ and $P'$, arranged so that each of the three shaded sectors, $SPP'$, has the same area. The second law states that since each of these areas is the same, the planet must move through each of the corresponding arcs, $PP'$, in equal times. When near the sun the planet must move relatively quickly so that the short line $SP$ will sweep out the same area per unit time as is swept out by the longer line $SP$ when the planet is moving more slowly farther from the sun.

■ Reprinted by permission of the publisher from *The Copernican Revolution: Planetary Astronomy in the Development of Western Thought* by Thomas S. Kuhn, p. 213, Cambridge, Mass.: Harvard University Press, Copyright © 1957 by the President and Fellows of Harvard College, Copyright © renewed 1985 by Thomas S. Kuhn.

a cause, Kepler suggested that the sun exerts a force on the planets that pushes them around their orbits. In the first edition of the *Mysterium cosmographicum*, he called this force *anima motrix*, or "motive soul." Twenty-five years later, when he was preparing a second edition of his book, he changed the term to *vis motrix*, or "motive force." He changed both the concept and the terminology because he realized that "this motive force grows weaker as the distance from the sun increases, just as the light of the sun is attenuated."[7]

Because of his speculations about a physical force connecting the sun to the planets, Kepler continued to search for a mathematical relationship between the motions of the planets and their respective distances from the sun. Eventually he discovered the relationship that later became known as his third law, that the square of the planet's period (the time it takes to complete an orbit around the sun) is proportional to the cube of its mean distance from the sun:

$$T^2 \propto r^3,$$

where $T$ is the planet's period and $r$ represents its mean distance from the sun.

Kepler published this result, along with other relationships revealing the pure harmonies which he thought guided God in the work of creation in *Harmonice mundi* (*Harmony of the World*) (1618), a book intended to complete the project begun in the *Mysterium cosmographicum*. These works contained speculations about musical harmonies produced by the motions of the planets and the geometrical architecture of the universe, as well as speculations about the forces causing the motions of the heavenly bodies. Pythagorean and theological assumptions permeated his astronomical thinking.

Although Kepler did not isolate and highlight the results that later thinkers identified as Kepler's laws, his work in astronomy represented a truly revolutionary break from the past. Unlike Copernicus and Tycho, he cast aside the ancient assumption of uniform circular motion, replacing circles with ellipses, and removed the sun from the geometrical center of the solar system. He unequivocally asserted the physical reality of his astronomical theory and attempted to provide a physical explanation for the motions of the planets. Instead of explaining the heavenly motions in terms of Aristotelian forms, he found mathematical relationships and the geometrical structure of the universe deeply explanatory. He fused mathematics, astronomy, physics, and theology

7. Quoted by Richard S. Westfall in *The Construction of Modern Science: Mechanisms and Mechanics* (New York: John Wiley & Sons, 1971), pp. 9–10.

into an integrated system to explain the universe. In so doing, he radically al-tered the traditional relationships among these disciplines.

Kepler's astronomy raised new problems, even as it provided a sound answer to the ancient problem of the planets. In a cosmology based on Aristotelian assumptions, circular motions around a center, as the natural motion of heavenly bodies made of quintessence, required no special explanation. Motion in an ellipse around one focus rather than around the center did not share the same natural characteristics. It called for a physical explanation. After Tycho's elimination of the solid spheres and Aristotelian natural motion, Kepler left astronomers and natural philosophers two urgent questions: What holds the planets in their orbits? And how can the elliptical orbits be explained? Answers to these questions awaited fundamental changes in the science of motion.

The implications of the new astronomy were profound. The Copernican challenge to geocentrism in astronomy challenged Aristotle's theory of motion, his theory of matter, and ultimately the metaphysics that underpinned all of these theories. In addition to introducing major changes into mathematical astronomy, Copernican astronomy increased the need for a new physics and a new philosophy of nature.

Each of these developments—Renaissance humanism, the Protestant Reformation, the discovery and exploration of the New World, and Copernican astronomy—contributed to the erosion of the worldview that had held sway for close to two thousand years. They undermined the authority of Aristotle, Galen, and Ptolemy, and led scholars and natural philosophers to search for a new philosophy of nature that would provide foundations for their knowledge of the world. Ironically, a backward-looking Renaissance humanism created a fertile matrix for innovation in the sciences.

# 3 Observing the Heavens

## From Aristotelian Cosmology to the Uniformity of Nature

The revolutionary developments in mathematical astronomy and the consequent challenge to Aristotelian physics and cosmology raised urgent new questions. If the heavens are no longer filled with solid spheres, what is the nature of space? What holds the planets in their orbits? Are the heavens qualitatively different from the region around the Earth? Two developments addressed aspects of these questions. The introduction of the telescope led to the discovery of previously unknown phenomena, many of which continued to erode Aristotelian cosmology. And a fundamental reconceptualization of motion answered the objections to the earth's motion and ultimately led to an explanation of the motions of the planets. Galileo Galilei (1564–1642) made important contributions to both areas.

### Gazing at the Stars: Galileo and the Telescope

Although lenses had long served as magnifying glasses, as burning lenses to start fires by concentrating the heat of the sun's rays on a single spot, and as spectacles to improve vision, no one before the seventeenth century had thought to use a convex lens as an objective and a concave lens as an eyepiece to make distant objects appear larger. The first reports of the invention of such an instrument surfaced in the Dutch Republic in 1608 as the work of the spectacle maker Hans Lipperhey (d. 1619), who constructed a three- or four-powered telescope. Word of this invention spread rapidly through Europe, and several people interested in astronomy—including Thomas Harriot (ca. 1560–1621) in England and Galileo Galilei (1564–1642) in Italy, a professor of mathematics at the University of Padua—assembled telescopes and used them to observe the heavens. Galileo had begun his career as a professor of mathematics at the University of Pisa, where he taught from 1589 to 1590 before moving to the University of Padua, where he remained until 1610.

By August 1609, Galileo was using a telescope of eight powers of magnification. By the end of the year, Galileo had improved the telescope to twenty powers, and he continued to improve it over the next few months until he could see things thirty times the size that they appear to the naked eye. Pointing the new instrument to the heavens, he made a number of astonishing observations, an account of which he published in the *Sidereus nuncius* (*The Sidereal Messenger*) in 1610. In this short work, he described phenomena never before observed. Looking at the moon, he noticed that "the boundary dividing the bright from the dark part does not form a uniformly oval line, as would happen in a perfectly spherical solid, but is marked by an uneven, rough, and very sinuous line."[1] Furthermore, there were bright spots in the darkened part of the moon and dark spots in the bright part. He interpreted these observations as indicating that the moon has a rough surface, parts of which are mountainous. The moon, he realized, resembles the earth. It is not a perfectly smooth sphere as Aristotle had claimed but, like the earth, has a rough and uneven mountainous surface.

Galileo observed that the dark part of the crescent moon is often dimly illuminated. He explained this phenomenon as earthshine. Just as light from the moon can illuminate a dark night on earth, so light reflected from the earth can illuminate the part of the moon which does not reflect direct light from the sun. These observations further supported the claim that the earth and moon are similar kinds of objects and directly challenged the Aristotelian view that celestial bodies are different in kind from terrestrial objects.

Galileo's telescope resolved the appearance of the Milky Way into numerous individual stars, thus settling a long-standing controversy about the nature of the galaxy. The telescope also revealed that Venus, like the moon, has a full set of phases. The Ptolemaic system can explain only some of the phases, but Copernican cosmology can explain the full set very easily.

Using the telescope, Galileo made the next dramatic discovery, the satellites of Jupiter. Observing Jupiter over many nights, Galileo was perplexed by what he saw. The planet appeared to be moving back and forth against the background of fixed stars. At first Galileo thought that Jupiter might be undergoing direct motion even though calculations indicated that it should be moving in the retrograde direction. After making many observations, he concluded that the motions were due to these "stars" and not to Jupiter: "Since I knew . . . that the observed stars were always the same ones (for no others, either preceding or following Jupiter, were present along the Zodiac for a great distance), now, moving from doubt to astonishment, I found that the observed change was not in Jupiter but in the said stars." Continuing to observe these

1. Galileo Galilei, *Sidereus Nuncius; or, The Sidereal Messenger*, trans. Albert Van Helden (Chicago: University of Chicago Press, 1989), p. 40.

anomalous stars, "I therefore arrived at the conclusion, entirely beyond doubt, that in the heavens there are three stars wandering around Jupiter like Venus and Mercury around the sun. This was at length seen clear as day in many subsequent observations, and also that there are not only three, but four wandering stars making their revolutions about Jupiter."[2] The startling discovery that Jupiter has satellites added support for the Copernican argument that the earth is a planet like all the others, for the satellites of Jupiter displaced the earth from its special status as the only planet having a moon. Galileo called the moons of Jupiter the Medicean Stars, after the rich Florentine banking family, the Medici, from whom he was at this time seeking patronage.

The *Sidereus nuncius* provoked various responses, reflecting the profound questions that Galileo's telescopic observations raised, not only for astronomers, but for natural philosophers and theologians as well. The most enthusiastic response came from Kepler, who published *Dissertatio cum nuncio sidereo* (*Conversation with the Sidereal Messenger*) in April 1610, barely a month after Galileo's book had appeared. Using telescopes available at the court of Rudolph II in Prague, Kepler replicated Galileo's observations of the moon, but none of the telescopes available to him was sufficiently powerful to reveal the satellites of Jupiter. Although Kepler's *Dissertatio*—reprinted and circulated widely through Europe—gave Galileo important support by justifying the use of the telescope to study the heavens, Galileo never responded to Kepler and never thanked him for the support.

Not everyone shared Kepler's enthusiasm. Cesare Cremonini (1550–1631), Galileo's friend and colleague at the University of Padua, rejected the telescopic observations out of hand. As reactionary and dogmatic as Cremonini's attitude may appear, it reveals important problems concerning the introduction of a new instrument. Remember that the Aristotelian cosmos was divided into two regions—terrestrial and celestial—each containing its own kind of matter and each governed by its own laws of nature. Although Galileo had tested the veracity of the telescope on the surface of the earth, he had no way of verifying that it produced accurate images of celestial objects. Indeed, the allegedly different composition of the heavens and the terrestrial region would suggest that using this earth-bound instrument could not possibly reveal anything about the heavens. How can we know that light coming from the heavenly bodies has the same properties as light near the surface of the earth? How can we know that the telescope does not create illusory images, like the chromatic distortion that appears at the edges of the visual field? In light of these theoretical objec-

2. Ibid., p. 66.

tions to the telescope, many Aristotelians considered Galileo's observations to be illusions created by the instrument. To assert that the telescope could make valid observations of heavenly bodies amounted to asserting that the celestial and terrestrial regions consist of the same kind of matter, the very point that Galileo had set out to prove.

Attempting to validate his instrument and his results to other astronomers, Galileo visited the astronomer Giovanni Antonio Magini (1555–1617) in Bologna, where he set up his telescope in hopes of demonstrating the most dramatic of his observations. The attempt failed. According to the report of Magini's young Bohemian associate, Martin Horky, they tried to test Galileo's telescope, but the observations made were clearly deceptive. Individual stars appeared double, thus casting into doubt Galileo's claim that Jupiter has satellites. In a letter describing the trials, Horky wrote to Kepler, "On earth [the telescope] works miracles; in the heavens it deceives."[3] This failure served only to reinforce the conservative skeptics.

A far more positive response came from the astronomers at the Collegio Romano, a Jesuit university in Rome. Cardinal Robert Bellarmine (1542–1621), a prominent and learned theologian who served as head of the Collegio Romano, asked his mathematicians to answer the following questions about Galileo's reported discoveries:

> First, whether you confirm the multitude of fixed stars invisible with the naked eye, and in particular that the Milky Way and the nebulosities are a congeries of very small stars.
>
> 1. That Saturn is not a simple star but three stars joined together.
> 2. That the star of Venus changes its shape, waxing and waning like the moon.
> 3. That the moon has a rough and uneven surface.
> 4. That about four movable stars go around the planet of Jupiter, and with motions different among themselves and very swift.[4]

The mathematicians confirmed the accuracy of the observations. Nevertheless, they had a few qualifications. For example, they noted: "We have observed that Saturn is not circular, as Jupiter and Mars appear, but is of an oval figure, like this oOo, although we have not seen the two starlets on either side sufficiently separated from the one in the middle to be able to say that they are distinct stars."[5] They confirmed the significant observations of the rough

3. Quoted ibid., p. 93.
4. Quoted ibid., p. 110.
5. Quoted ibid., p. 111.

texture of the moon and the satellites of Jupiter. Although they did not agree with all of Galileo's interpretations of the observations, they did validate the telescope as a legitimate tool for observing the heavens.

Pointing the telescope to the heavens, Galileo discovered previously unknown phenomena that were not compatible with either Ptolemaic astronomy or Aristotelian cosmology. His observations, however, did not provide direct proof for Copernicanism because Tycho's system could explain most of the observed phenomena equally well. Galileo's telescopic observations provided powerful evidence that the region of the heavens is made of the same sort of matter as the earth. This assertion of the spatial uniformity of nature directly contradicted the Aristotelian claim that the cosmos consists of two distinct regions—the terrestrial and the celestial—that are composed of different kinds of matter and are governed by different laws, thus adding to the factors leading to the demise of Aristotelian cosmology.

## Interpreting the Bible: Galileo and the Church

The exciting discoveries recounted in the *Siderius nuncius* boosted Galileo's public reputation dramatically. Publication of the book and his astuteness in naming the moons of Jupiter the Medicean Stars secured the patronage of the wealthy Florentine Medici family, who appointed him philosopher and chief mathematician to the grand duke of Tuscany, a position he held from 1610. This appointment increased his social status, freed him from the duties of university teaching, and provided him with a significantly increased income. Galileo received a very different kind of recognition as well. In 1611, he received an invitation to join the tiny Accademia dei Lincei (Academy of the Lynxes), founded by the young nobleman, and later prince, Frederico Cesi (1585–1630). Cesi had formed the group to pursue natural history and natural philosophy. Galileo was proud of his membership in this proto-scientific society, which supported his publication on sunspots in 1613 and his role in the controversies that followed for the next ten years. The Accademia dei Lincei survived Cesi's early death and lasted until Galileo's condemnation in 1633, after which it ceased to exist.

Despite the Jesuit astronomers' willingness to accept Galileo's new observations, their attitude did not represent that of the entire Catholic Church. Several Dominican friars who considered Galileo's new cosmology to be incompatible with sacred Scripture openly opposed him. Since the Dominicans ran the Inquisition, their early attacks on Galileo foreshadowed future problems. As early as 1612, the Dominican Nicccolò Lorini publicly attacked Copernican ideas as contrary to Scripture. Ignorant about these matters, however, he referred to "Ipernicus, or whatever his name is."[6] Lorini later filed a complaint against Galileo to the Inquisition.

In 1613 Benedetto Castelli, one of Galileo's students, joined the faculty of the University of Pisa. He wrote a letter to Galileo describing a breakfast with the Dowager Duchess during which the topic of Galileo's telescopic discoveries arose. Castelli reported that the Grand Duchess asked him about the religious implications of Copernican astronomy. Galileo replied in a letter to Castelli, in which he outlined his views on the proper relationship between science and Scripture, stating that since science is based on sense experience and necessary demonstration, it should not be questioned in cases where it disagrees with statements in Scripture. Rather (borrowing an approach from St. Augustine), he stated that, since the Bible was written in a style designed to be understood by ordinary people, the passages in question could be reinterpreted in non-literal ways.

In 1614, another Dominican, Tommaso Caccini, delivered a sermon in Florence on a text from the book of the Acts of the Apostles (1:11), the Latin version of which reads, "Viri Galilaei, quid statis adspicientes in coelum?" ("Men of Galilee, what are you looking for in the sky?")[7] Clearly, Galileo had the signal honor of being mentioned by name in the Bible! In the meantime, Lorini obtained a copy of Galileo's letter to Castelli and sent it to the prefect of the Congregation of the Index of Forbidden Books in Rome, an organization that determined which books to ban for containing statements contrary to orthodox doctrine.

Galileo realized that powerful forces opposed his work and decided to rewrite the letter to Castelli in a form intended for public consumption. Accordingly, he published his definitive statement on the issue, the *Letter to Madame Christina of Lorraine, Grand Duchess of Tuscany, Concerning the Use of Biblical Quotations in Matters of Science* (1615). He acknowledged that a number of biblical passages, when interpreted literally, seem to contradict Copernican cosmology—most explicitly in the book of Joshua. There, a passage states that God ordered the sun to stand still in order to prolong the hours of daylight, thereby giving Joshua enough time win a battle. Galileo argued that the Bible cannot say something false when we understand its meaning correctly but that it often speaks in figurative language, as when it refers to the hands and feet of God. These figurative expressions accommodate the text to the understanding

6. Quoted in Annibale Fantoli, *Galileo: For Copernicanism and for the Church*, trans. George V. Coyne, 2nd ed. (Rome: Vatican Observatory Publications, 1996), p. 171.
7. Ibid., p. 175.

of ordinary people. Invoking the metaphor of the two books—the book of God's word and the book of God's work—Galileo stated, "This being granted, I think that in discussions of physical problems we ought to begin not from the authority of scriptural passages, but from sense-experiences and necessary demonstrations; for the holy Bible and the phenomena of nature proceed alike from the divine Word, the former as the dictate of the Holy Ghost and the latter as the observant executrix of God's commands."[8] The constancy and uniformity of nature entail the reliability of sense experience. The Bible, on the other hand, can be interpreted figuratively. Although we should seek truth in Scripture, the truth we find there is spiritual truth. God would not have given us senses and reason if he did not intend us to use them. In the words of Cardinal Cesare Baronio (1538–1607)—whom Galileo called "an ecclesiastic of the most eminent degree"—"the intention of the Holy Ghost is to teach us how one goes to heaven, not how heaven goes."[9] Galileo concluded the *Letter to the Grand Duchess* with a Copernican interpretation of the miracle of Joshua. Sense and faith, he declared, do not contradict each other, but, properly interpreted, are mutually supportive.

The question of scriptural interpretation was highly charged in post-Reformation Europe. At the Council of Trent, the Roman Catholic Church had reasserted its own authority as the sole interpreter of Scripture, in accordance with the unanimous agreement of the Church Fathers, the theologians writing during the first few centuries of Christianity. Galileo's *Letter to the Grand Duchess* led to trouble for three reasons. First, Galileo naively assumed that he could undertake the task of interpretation simply because the position he took seemed reasonable to him. He seems to have been blind to the controversial nature of biblical exegesis in the wake of the Reformation and the decrees of the Council of Trent. Second, his support of Copernicanism contradicted the Church's traditional geocentric and geostatic cosmology, which found support in both the Bible and Aristotelianism. And third, Galileo used the principle of accommodation—the claim that portions of the Bible were written so that they were accommodated to the understanding of ordinary people—which was an exegetical strategy often invoked by the reformers and one that consequently seemed provocative to the Church authorities, despite the fact that this approach originated with St. Augustine.

8. Galileo Galileo, *Letter to Madame Christina of Lorraine, Grand Duchess of Tuscany, Concerning the Use of Biblical Quotations in Matters of Science*, in Stillman Drake, *Discoveries and Opinions of Galileo* (Garden City, NY: Doubleday Anchor, 1957), p. 182.
9. Ibid., p. 186.

Shortly after Galileo published the *Letter to the Grand Duchess*, he got wind of the fact that his enemies, particularly Caccini, were seeking to have his work censured by the Inquisition. Partly in response to these rumors, he spent the winter of 1615–16 in Rome, aggressively arguing in favor of Copernican astronomy. His persistence provoked the Church to respond. At Bellarmine's request, a group of theologians who worked within the bureaucracy of the Inquisition examined the two fundamental propositions of the Copernican system:

1. The sun is the center of the world and hence immovable of local motion.
2. The earth is not the center of the world, nor immovable, but moves according to the whole of itself, also with a diurnal motion.[10]

They made the following judgments about these statements.

On the first:

All said that this proposition is foolish and absurd in philosophy [i.e., physics], and formally heretical since it explicitly contradicts in many places the sense of the Holy Scripture, to the common interpretation and understanding of the Holy Fathers and the doctors of theology.

On the second:

All said that this proposition receives the same censure [qualification] in philosophy and that in regard to theological truth it is at least erroneous in faith.[11]

The cardinals of the Holy Office (the Inquisition) considered these judgments and asked Bellarmine, now elevated to cardinal, "to call Galileo before himself and warn him to abandon these opinions; and if he should refuse to obey, the Father Commissary [head of the Holy Office], in the presence of notary and witnesses, is to issue him an injunction to abstain completely from teaching or defending this doctrine and opinion or from discussing it; and further, if he should not acquiesce, he is to be imprisoned."[12]

Although some ambiguity remains in the records about whether or not the Father Commissary jumped the gun and gave Galileo the stronger warning before he had a chance to respond to Bellarmine, we do know that Galileo received Bellarmine's admonition, and for many years ceased writing directly about Copernicanism. Following this hearing Copernicus' *De revolutionibus*

10. Quoted in Fantoli, *Galileo*, p. 215.
11. Ibid., p. 216.
12. Ibid., pp. 217–18.

and Kepler's books were placed on the Index of Forbidden Books (until corrected), where they remained until 1757.

Bellarmine's position on the Copernican theory is evident from his reaction to the case of Paolo Antonio Foscarini (1580?–1616), a Carmelite monk and teacher of theology, who had earlier been imprisoned for endorsing Copernicanism and recommending the reinterpretation Scripture along lines similar to those suggested by Galileo. Bellarmine wrote a long letter to Foscarini, outlining his position on these questions. He said that he found no problem with supporting Copernicanism as a useful hypothesis but heartily objected to the claim that the theory describes reality because "this is a very dangerous thing, likely not only to irritate all scholastic philosophers and theologians, but also to harm the Holy Faith by rendering Holy Scripture false."[13] Bellarmine reminded Foscarini of the finding of the Council of Trent, which "prohibits interpreting Scripture against the common consensus of the Holy Fathers."[14] Finally, he said: "If there were a true demonstration that the sun is at the center of the world and the earth in the third heaven, and that the sun does not circle the earth but the earth circles the sun, then one would have to proceed with great care in explaining the Scriptures that appear contrary; and say rather that we do not understand them than that what is demonstrated is false. But I will not believe that there is such a demonstration, until it is shown me."[15]

Bellarmine knew, and so did Galileo, that visual data alone could not prove the reality of heliocentrism. Such a proof required physical evidence, and no such evidence existed. Galileo's new science of motion answered the objections to the earth's motion, but it did not provide positive proof that the earth moves. The telescopic evidence had convinced the Jesuits, including Bellarmine, that Aristotelian cosmology and Ptolemaic astronomy were no longer viable theories, but the Jesuits opted for Tycho's system because it maintained a stationary earth about which the sun revolves.

In the years following his hearing with Bellarmine, Galileo continued to write about various topics in mathematics and natural philosophy, but he did not deal directly with Copernicanism. In 1623, Pope Paul V died, and Cardinal Maffeo Barberini (1568–1644) became pope and assumed the name Urban VIII. Barberini supported the new developments in the sciences. Although he still considered all astronomical theories to be hypotheses, he had intervened to prevent Copernicanism from being declared heretical at the time of Galileo's troubles in 1616. Galileo welcomed Barberini's election as encouraging news. Adding to his optimism was the fact that the pope granted him six hour-long meetings in April 1624. Although the pope did not change his view about the hypothetical status of astronomical theories, Galileo left feeling free once again to discuss Copernicanism, as long as he dealt with the theory hypothetically.

   With this condition in mind, he set out to write his great summary of the Copernican debates, the *Dialogue Concerning the Two Chief World Systems— Ptolemaic and Copernican* (1632). The *Dialogue* describes four days of discussions about the relative merits of Aristotelian cosmology and Ptolemaic astronomy, on the one hand, and Copernican astronomy, on the other. The three interlocutors are Salviati, who takes Galileo's position; Simplicio (whose name resonates with that of a sixth-century Aristotelian commentator, Simplicius), who defends the traditional views; and Sagredo, an educated layman whose opinion both of the others try to sway. Although Galileo used the form of the dialogue as a way of dealing with the astronomical theories hypothetically, Salviati refutes Simplicio's arguments so overwhelmingly that the pretense of hypothesis is a transparent ruse.

   On the first day of the *Dialogue*, the interlocutors discuss the fundamental uniformity between the celestial and terrestrial regions in opposition to Aristotelian cosmology. Salviati appeals to Tycho's observations of comets and new stars to demonstrate the mutability of the heavens and to Galileo's telescopic observations to note the physical similarity between the two regions. He also describes Galileo's new science of motion to answer the Aristotelian objections to the earth's motion. During the second day, discussion of the earth's motion continues, focusing on its daily rotation on its axis. On the third day, the interlocutors consider the earth's annual revolution around the sun, and Salviati once again appeals to various observations to demonstrate the physical possibility of this motion. The fourth day is the climax of the debate, in which Salviati presents a theory of the tides that Galileo believed was proof positive of the earth's motion. Despite the deep flaws in his theory, Galileo repeatedly invoked it, because even he realized that all his other arguments at most demonstrated the possibility of the earth's motion but did not prove that it actually moves.

   Galileo completed the book sometime in 1629. All books published in Church jurisdictions at this time needed approval from ecclesiastical censors before they could be published. In 1630, Galileo submitted the *Dialogue* to the censor in Rome, who determined that it was insufficiently hypothetical and that it needed some revisions to address that problem. The censor called for a new preface to note these revisions. Meanwhile, Galileo had returned home to Florence, which was under quarantine because of an outbreak of plague. Obtaining permission from Rome to submit the book to a censor in Florence, he finally received the censors' approval to publish. The book appeared in 1632.

Shortly after the *Dialogue* was published, Galileo's enemies started to move against him and his troubles escalated. The Jesuits, whose support Galileo had lost years before in an overheated dispute about the sunspots, lobbied the pope to ban the book. The pope became hostile to the book as he became aware of its nature. The preface that censors had demanded was separate from the body of the book and therefore appeared to be extraneous. Furthermore, the Aristotelian interlocutor Simplicio, who came off as a fool (an impression reinforced by Simplicio's name, which means "simpleton"), expressed some of the views that the pope had communicated to Galileo during their earlier meetings. The censor in Rome, acting under instructions from the pope, sought to recall all the books in circulation. Meantime, the Holy Office had discovered the minutes of Galileo's hearing with Bellarmine in 1616, in which the inquisitors found the injunction that the Commissary General of the Inquisition had supposedly delivered to Galileo—namely, that he was "to abstain completely from teaching or defending this doctrine and opinion or from discussing it; and further, if he should not acquiesce, he is to be imprisoned."[16]

The Inquisition summoned Galileo and indicted him on three counts: first, that he had transgressed the previous orders by deviating from hypothetical treatment of Copernicanism; second, that he had erroneously ascribed the phenomenon of the tides to the stability of the sun and the motion of the earth, motions which do not exist; and third, that he had been deceitfully silent about the command laid on him by the Holy Office in 1616. The case came to trial in 1633. By this time, Galileo was the only person still living who had been present at the hearing in 1616, and the minutes of that hearing—which the Inquisition had probably altered—were the crucial evidence against him. In the end, Galileo was forced to recant, but refused, even under the threat of torture, to admit malicious intent to be deceitfully silent about the order of 1616. He received a sentence of life in prison, but the sentence was commuted to life under house arrest. He never endured torture, and he did not spend time languishing in prison. The Church banned the *Dialogue*, and it remained on the Index of Forbidden Books until the nineteenth century.

Until his death in 1642, Galileo continued working. He completed the *Dis-*

16. Ibid., pp. 217–18.

*courses on Two New Sciences,* his book on the science of motion and the strength of materials. Some of his students smuggled that book out of Italy and took it to Amsterdam, which lay beyond the reach of the Church. Published in 1638, the book played a critical role in the formation of a new science of motion in the seventeenth century.

## Using Astronomy: The Effects of the Stars

In addition to enlarging their stock of observations and refining their mathematical theories, early modern astronomers and natural philosophers remained concerned with the effects of the heavenly bodies on terrestrial events. Although today we distinguish between astronomy (as the study of the motions and properties of heavenly bodies) and astrology (as the theory of the influence of heavenly bodies on terrestrial events and human lives), the two disciplines were not clearly distinguished in early modern times. Indeed, the descriptions of courses at some Italian universities used the two words interchangeably well into the eighteenth century.

Early modern thinkers distinguished between two aspects of astrology: natural astrology dealt with the effects of the heavens on natural phenomena such as climate, agriculture, and collective human events like epidemics or major political or religious movements; and judicial astrology, based on natal horoscopes, dealt with the effect of the heavenly bodies on the course of individual lives. No one doubted that the motions of the sun and the moon directly affect the growth of crops and the regular course of the seasons. Similarly, they argued, the motions of the heavenly bodies could affect human health, directly or indirectly. In explaining these effects, astrologers appealed to natural causes as described in the prevailing natural philosophy.

Controversy perennially engulfed judicial astrology, from its beginnings in ancient times well into the seventeenth century. From the time of St. Augustine in the fourth century, Christian theologians denounced astrology because the fatal effect of the stars would limit both divine and human free will. Thomas Aquinas, in the thirteenth century, accepted the influence of the stars on human life because, he argued, the body can affect the mind, and the stars can affect the body. However, people can resist their passions by exerting their free will and thus resist the effects of the stars. Quoting Ptolemy, he said, the stars incline but do not compel.

The Renaissance Hermetic philosopher Giovanni Pico della Mirandola (ca. 1463–94) wrote an extended attack on astrology, *Disputationes adversus astrologiam divinatricem* (*Disputations against Divinatory Astrology*) (1495). The fact that a contemporary astrologer had predicted his early death may well have motivated his argument. Pico distinguished astrology from astronomy: "When I say astrology I do not mean the mathematical measurement of stellar sizes and motions, which is an exact and noble art . . . but the reading of forecoming events by the stars, which is a cheat of mercenary liars, prohibited by both civil and church law, preserved by human curiosity, mocked by philosophers, cultivated by itinerant hawkers, and suspect to the best and most prudent men."[17]

Continuing his diatribe, Pico called astrology "the most infectious of all frauds, since, as we shall show, it corrupts all philosophy, falsifies medicine, weakens religion, begets or strengthens superstition, encourages idolatry, destroys prudence, pollutes morality, defames heaven, and makes men unhappy, troubled, and uneasy; instead of free, servile and unsuccessful in nearly all their undertakings."[18] He noted that astrology produces uncertain results. As both Arabic and Hebrew philosophers had said, predictions may fail "[to] come true because the matter is not suited to receive influence, or free will can intervene, or the Divine purpose may 'ordain things otherwise than the usual revolution of the heavens would effect.'"[19] He argued against some of the technical aspects of astrology, especially ideas that he considered arbitrary, such as the notion of 360 degrees in a circle or that parts of empty space could affect the lives of individuals. Pico argued that the heavenly bodies can act by means of light and heat, but not by the more occult influences and correspondences claimed by some advocates of judicial astrology.

Pico's attack on astrology led others to defend the art. Lucio Bellanti (d. 1499), the astrologer who had (correctly, as it happened) predicted Pico's early death, wrote a long rebuttal to Pico's attacks on the art, defending the status of astrology as a legitimate Aristotelian science. Giovanni Pontano (1422?–1503) also wrote a reply to Pico's denunciation of astrology. Pontano agreed that "the stars do not build ships, or supply the axes by which criminals are punished, or settle kingdoms or prefectures, all these activities being determined by men."[20] But by acting on the humors from which people are made, they impress their

17. Giovanni Pico della Mirandola, *Disputationes adversus astrologiam divinatricem*, ed. Eugenio Garin, 2 vols. (Florence, Italy: Vallecchi, 1946–52), 1:40; trans. and quoted in Wayne Shumaker, *The Occult Sciences in the Renaissance: A Study in Intellectual Patterns* (Berkeley and Los Angeles: University of California Press, 1972), pp. 18–19.

18. Pico, *Disputationes*, 1:44 (Shumaker, p. 19).

19. Pico, *Disputationes*, 1:100–102 (Shumaker, p. 19).

20. Pontano, quoted in Shumaker, *Occult Sciences*, p. 31.

effects on individuals. Although local circumstances will determine the specific ways outcomes are achieved, astrological influences will nevertheless determine the general outlines of events such as death, success, or illness.

The Protestant reformers had mixed views on astrology. Philip Melancthon (1497–1560), a leading Lutheran theologian, argued that God's providence guarantees that every part of the creation has a purpose. Consequently, the heavenly bodies provide knowledge of human nature and human affairs. Jean Calvin, on the other hand, rejected judicial astrology on both theological and methodological grounds. He argued that divine freedom allows God to overcome the influence of the heavens and that people are often profoundly changed by the experience of conversion. He thought that astrology leads to ridiculous conclusions. For example, even though twins share the same horoscope, often their lives turn out very differently. He asked whether the thousands of men who die in a single battle must share the same horoscope. He carried this argument to the extreme in demonstrating what he considered the absurdity of judicial astrology. The astrologers

> will tell a man how many wives he will have. Yes—but do they find in his star the horoscope of his first wife, so that they know how long she will live? By this process the wives will be made to have no horoscopes of their own. . . . In brief, by this reasoning the horoscope of every individual man will include a judgement of the whole disposition of a country, since [the astrologers] boast that they can judge whether a man will be happy in marriage, whether he will have fortunate or unfortunate meetings with other people in the fields, what dangers he may fall into, whether he will be killed or will die of a disease. Consider with how many people we have commerce during our lives.[21]

Renaissance debates about astrology parallel modern controversies about the roles of heredity and environment in human development.

Most of the major early modern astronomers did cast horoscopes, often to meet their patrons' demands for almanacs, nativities, and prognostications. For example, Tycho Brahe, combining his commitments to Lutheran theology and Paracelsian cosmology (based on a correspondence between the macrocosm, or the heavens, and the microcosm, or the human being), justified judicial astrology on the grounds that no part of the creation lacks a providential purpose and that the purpose of the heavens is to provide insight into

---

21. Jean Calvin, *Avertissement contre l'astrologie: Traité des reliques; suivies du Discours de Théodore de Bèze sur la vie et la mort* (Paris: Librairie Armand Colin, 1962), p. 11 (trans. Shumaker, *Occult Sciences*, p. 45).

the microcosm of the human being. Kepler cast horoscopes reluctantly, not because he disapproved of astrology, but because he thought that some of its methods needed revision. Interestingly, the rise of Copernican astronomy did not undermine judicial astrology, as horoscopes continued to be drawn from a geocentric standpoint: the position of the heavenly bodies relative to the earth is the astrologically significant variable. Although interest in astrology continued in some circles and especially in popular culture, by the end of the seventeenth century astronomers and natural philosophers no longer cast horoscopes or expressed much interest in judicial astrology.

Comets had played a special role in astrological thinking. Because of their apparently unpredictable appearance, people interpreted them as portents of political and religious events on earth. For example, English astrologers interpreted the great comet of the winter of 1680–81 as a portent of dire consequences following from the political crisis of 1679–81—the result of the attempt by the English king Charles II, who was heirless, to ban his Catholic brother from inheriting the throne. But, as the astronomer John Flamsteed (1646–1719) had written a few years earlier, if the comet of 1677 happened to return in twelve years' time, "it will wholly overthrow ye conjectures [and] fearfull predictions of ye Astrologers."[22] Within a few years, Newton was able to prove that many comets do indeed return at regular intervals. In fact, like the planets, comets follow elliptical orbits about the sun at one focus, although their orbits are so eccentric that they can be treated as parabolas. This result received dramatic reinforcement when the astronomer Edmond Halley (ca. 1656–1743) demonstrated that the comet of 1681 was the same comet that had previously appeared at intervals of about 75 years.

The discovery that comets are orbiting bodies that reappear at regular intervals dealt a serious blow to their astrological meaning. As Halley wrote in his "Ode to Newton," which served as a preface to the *Principia*,

Now we know what curved path the frightful comets have;
No longer do we marvel at the appearances of a bearded star.[23]

22. Flamsteed to Richard Towneley, 11 May 1677, Royal Society MSS. LIX.c.10, quoted by Simon Schaffer, "Newton's Comets and the Transformation of Astrology," in *Astrology, Science, and Society: Historical Essays*, ed. Patrick Curry (Wolfeboro, NH: Boydell, 1987), p. 222.
    23. Edmond Halley, "Ode on This Splendid Ornament of Our Time and Our Nation, the Mathematico-Physical Treatise by the Eminent Newton," in Isaac Newton, *The Principia: Mathematical Principles of Natural Philosophy. A New Translation*, trans. I. Bernard Cohen and Anne Whitman, assisted by Julia Budenz (Berkeley and Los Angeles: University of California Press, 1999), p. 379.

Although they may have undermined the astrological meaning of comets, Newton and his followers endowed them with religious significance, interpreting them as important components of God's providential design of the universe.

In the wake of Copernicus' and Kepler's endorsement of heliocentric astronomy, new observations, especially those that Galileo made with the telescope, continued to erode Aristotelian cosmology. Though not explicitly stated at the time, these observations reinforced a closer relationship between the previously separate disciplines of physics and astronomy. Insofar as the new observations chipped away at Aristotelian cosmology, they had serious implications for a theology that had been closely aligned with Aristotelianism. The renewed emphasis on observation became a criterion by which to evaluate not only astronomical theories, but also approaches to biblical exegesis and the claims of traditional astrology.

# 4 Creating a New Philosophy of Nature

During the early seventeenth century, natural philosophers deliberately sought a new philosophy of nature to replace the Aristotelianism that had served as the conceptual framework for natural philosophy and had dominated university curricula for centuries. In the wake of the Protestant Reformation, the development of heliocentric astronomy, the discovery of novelties in the heavens, and explorations of hitherto unknown parts of the world, many thinkers came to reject traditional ways of understanding the natural world. Renaissance humanists, in recovering a number of ancient philosophies, made alternatives to Aristotelianism available to scholars, who often turned to ancient texts to work out new ideas. Although natural philosophers considered a number of different candidates for replacing Aristotelianism, many adopted some version of what came to be known as the mechanical philosophy, a philosophy that explained all natural phenomena in terms of matter and motion.

The mechanical philosophy derived from the philosophy of Epicurus (341–271 BC), who, like other ancient philosophers, sought the key to the good life. He considered the good life to be one that maximizes pleasure and minimizes pain. Epicurus believed that the greatest sources of human unhappiness, apart from bodily pain, are fear of the gods and anxiety about punishment after death. To eliminate these causes of distress, he sought to explain all natural phenomena in purely naturalistic terms, the chance collisions of material atoms (microscopic, indivisible particles of matter) in empty space, thus eliminating the gods' interference in human lives. He claimed that the human soul, like everything else in the world, is material, composed of atoms that are exceedingly small and swift. According to Epicurus, the soul does not survive death. Without an afterlife, all fears of death and postmortem punishment vanish. Epicurus believed that an infinite number of atoms had always existed. Epicureanism, while not strictly atheistic, denied that the gods play a role in the natural or human worlds, thus ruling out both any kind of divine intervention in human life and any kind of providence in the world. Because of its reputation as atheistic and materialistic, Epicureanism fell into disrepute in antiquity and almost sank into oblivion during the Christian Middle Ages. The revival of Epicureanism in the early modern period followed the recovery and publication of the writings of Epicurus and his Roman disciple Lucretius during the fifteenth century.

## Mechanizing the Universe: Matter and Motion

Early modern treatises on the mechanical philosophy typically opened with an account of the first principles of this philosophy of nature—namely, theories of space, time, matter, motion, and cause. Once having spelled out these first principles, which amounted to establishing the ultimate terms of explanation for their new philosophy, mechanical philosophers proceeded to demonstrate how these fundamental terms could explain all the qualities of bodies and all the phenomena in the natural world.

During the first half of the seventeenth century, two natural philosophers, Pierre Gassendi (1592–1655) and René Descartes (1596–1650), published the first systematic and the most influential accounts of the mechanical philosophy. Although they differed on a number of specific topics, they shared the commitment to explaining everything in the world in terms of matter and motion.

Gassendi, a Catholic priest actively engaged with the community of natural philosophers in France, undertook a lifelong project to modify Epicurean atomism in order to make it compatible with Christian theology. Accordingly, he insisted on several modifications of Epicureanism: the existence of God; God's creation of a finite number of indivisible atoms that he endowed with motion; God's continuing providential relationship to the creation; free will (both human and divine); and the existence of an immaterial, immortal human soul, which, he claimed, God infuses into each individual at the moment of conception. He asserted that atoms, colliding in empty space, are the constituents of the physical world. In his massive *Syntagma philosophicum* (*Philosophical Treatise*), published posthumously in 1658, Gassendi attempted to explain all the qualities of matter and all the phenomena in the world in terms of atoms and the void. Gassendi wrote in the style of a Renaissance humanist, working out his own ideas in dialogue with ancient Greek and Roman thinkers.

Gassendi thought that the atoms move in empty space—the void. In claiming that the void exists, he differed from almost all other natural philosophers, who shared Aristotle's view that the cosmos is full of matter. According to Gassendi, a large extramundane void contains the universe that God created. This extramundane void is a boundless, incorporeal extension. In addition to this cosmic space, Gassendi believed that void space separates the atoms. He appealed to arguments first articulated by the ancient atomists to justify the existence of this interstitial or interparticulate void. Epicurus had argued that in order for atoms to move, empty spaces into which they can go must exist. Gassendi also borrowed arguments from the Greek physicist and engineer Hero of Alexandria (fl. 62 AD). Hero had drawn an analogy between the matter composing bodies and the properties of a heap of sand or wheat. Just as air or water separates the individual grains of sand or wheat, so small void spaces separate the particles composing bodies. The action of a bellows graphically illustrates the fact that air can be compressed and then rarefied. Explaining compression and rarefaction requires the existence of small bits of void between the atoms. Gassendi cited other phenomena to make the same point: the saturation of water with salt, the dissemination of dyes through water, the penetration of air and water by light, heat, and cold, all of which he thought were composed of particles of matter. Gassendi thought that a third kind of void exists, one that he called the *coacervatum* (heaped together) void. It does not exist naturally, but various manmade machines, such as pumps, siphons, and bellows, can produce it. A device can produce a long-lasting void space, as in the example to which Gassendi devoted by far the most attention, the space above the mercury in a barometer.

A mercury barometer consists of a glass tube that is closed at one end. It is filled with mercury and then inverted into a small bowl. The mercury in the tube then drops to a height of approximately 30 inches (76 centimeters), leaving an apparently empty space above it. Two questions about the barometer puzzled natural philosophers at this time: What causes the mercury to be suspended in the tube? And is the space above the mercury devoid of all matter? According to the traditional, Aristotelian explanation of the barometer, nature's *horror vacui* (horror of the vacuum) causes the column of mercury to remain suspended. Aristotelians thought that some kind of vapor or ether fills the space above the mercury because they rejected the possibility that a vacuum or void can exist in nature.

Rejecting the traditional explanation of these phenomena, Gassendi invoked purely mechanical terms—the pressure and resistance of the air. He based his explanations on the experiments of a number of his contemporaries, particularly Evangelista Torricelli (1608–47) and Blaise Pascal (1623–62). Pascal started at the base of a mountain, Puy de Dôme, where he measured the height of the column of mercury in a barometer. He then carried the barometer up the mountain and observed that as he climbed higher, the column of mercury fell. He concluded that the weight of the air suspends the mercury in the tube. He based this claim on the assumption that with increasing altitude, the quantity of air above him decreased, thus exerting less pressure on the column of mercury and resulting in the descent of the column. Pascal also claimed to have performed similar experiments under water, producing similar results: the deeper the barometer was placed in the water, the higher the column of mercury rose. He ascribed the increasing height of the mercury to the increasing pressure of the water as he descended. Like Pascal, Gassendi interpreted these experiments as establishing the fact that the mercury rises because of the heaviness of the air, not because of nature's fear of a vacuum.

Barometric experiments raised the additional question of whether the space above the mercury is in fact void. Gassendi argued that light, particles of heat and cold, magnetic particles, and the particles that flow from the earth to cause gravity all pass through the glass tube into the space above the mercury, assumptions that would seem to negate the absolute vacuity of that space. He noted, however, that the matter of light, heat, cold, or magnetism could not penetrate the space above the mercury unless that space contained some void. Otherwise, two bodies would occupy the same space simultaneously, a patent impossibility. He concluded that the barometric experiments can be interpreted without invoking the occult *horror vacui* and that they support the existence of the void, even if some particles of matter do enter the space. Other natural philosophers interpreted the experiments differently. The physicist and mathematician Gilles Personne de Roberval (1602–75) accepted the existence of void space above the mercury but rejected the explanatory role of the column of air. Descartes accepted that the column of air explained the changing height of the mercury but rejected the existence of the void. Pascal accepted both.

Having satisfied himself that void exists, Gassendi went on to ask what exists within the void. He answered that matter does. He began his discussion of matter by delineating the boundary between corporeal substance (matter) and incorporeal substances (God, the angels, demons, and the rational soul). Material things differ from incorporeal things because they are endowed with bulk or mass, by which he simply meant that they are tangible and capable of resisting other bodies. Atoms compose material things. Atoms are perfectly full, solid, hard, indivisible particles. They all consist of the same kind of matter. Their minute size makes it impossible to observe them directly, but various commonly observed phenomena lend support to their existence. Wind demonstrates that invisible matter can produce visible, physical effects. Paving stones and plowshares gradually wear away because of constant rubbing, even though individual acts of rubbing produce no discernible changes; it follows that they consist of very small particles of matter. Odors travel through air because tiny particles flow from the original body to the nose. Atoms contain no void and are therefore indivisible. The interstitial or interparticulate void separates them from one another.

What other properties do these atoms possess? In addition to the geometrical properties of size and shape, atoms possess two properties necessary for them to be the material principle of things: resistance (or solidity) and weight. Solidity or resistance (impenetrability) enables atoms to act by contact. Their resistance, or solidity, along with their magnitude, distinguishes them from immaterial mathematical points. The innate heaviness of atoms enables them to move. Unlike Epicurus, who had considered the propensity for motion to be innate, Gassendi claimed that God endows atoms with mobility and activity. Although God imposes motion on atoms, their motion persists perpetually. God created atoms at the beginning, and then he fashioned the first things he created from them. All subsequent generation and corruption and all change result from the motion, impact, and rearrangement of the original atoms. Although atoms fall below the threshold of sense, the recently invented microscope revealed an amazingly complicated world invisible to the senses. Viewed through the microscope, particles of flour appear complex, consisting of diversely shaped parts, and the tiny mite possesses a number of distinct organs. Although atoms come in a finite variety of shapes, they can produce the multitude of things in the world in the same combinatorial way that only twenty-two letters of the (Latin) alphabet can produce the full complexity of language.

In a world consisting of atoms, questions about causality became questions about how atoms interact. According to Gassendi, the activity of atoms lies in their motion. God, the first cause, instills motion and hence activity into the atoms by his command. All physical change results from the local motion and impact of atoms. Whereas Aristotle, in the *Physics,* had enumerated several kinds of change—growth, decay, generation, corruption, and qualitative change—Gassendi reduced them all to the motions of atoms. Atoms communicate their motions to each other by contact and collision. Consequently, impact is the primary agent of change in the physical world. Even if contact between the mover and the moved is not evident—for example, in the case of magnetic attraction or the transmission of heat from fire—contact does occur, at the invisible, atomic level. Like Aristotle, Gassendi claimed that there is no action-at-a-distance in the world.

Through his extensive discussion of atoms and the void, motion and change, Gassendi had formulated a new conceptual framework for natural philosophy: Of what kinds of entities does the world consist? Atoms and the void. By what means do these entities interact? By motion and collision. In answering these questions, he had replaced traditional Aristotelianism with the mechanical conceptual framework within which he thought natural philosophy should be formulated.

Gassendi's contemporary René Descartes also proposed a mechanical philosophy of nature. Unlike Gassendi, who approached philosophical questions by first considering the opinions of ancient philosophers, Descartes tried to derive his philosophy from principles that he considered to be self-evident and certain. Using the skeptical arguments that had so much currency at the time, he argued that all previous knowledge was uncertain and subject to doubt. After considerable reflection, he claimed that the statement "I think, therefore I am" cannot be doubted because the very act of doubting requires the existence of a doubting subject. He then defined mind as a thinking thing. He contrasted mind with matter, which he called "extended thing," arguing that the only property that matter continues to possess as it undergoes any kind of qualitative change is extension—that is, it occupies three dimensions. The impossibility of void space follows immediately from Descartes' identification of matter and extension. So does the infinite divisibility of matter, because the infinite divisibility of geometrical space entails the infinite divisibility of spatial extension.

Explaining variations in density proved difficult for Descartes because he thought that matter fills all space. Without bits of void interspersed among the material particles composing larger bodies, all bodies would have the same density. Empirically we observe that there are differences in density—say, between a feather and a piece of lead. Descartes tried to explain this fact, while adhering to his assumptions about the nature of matter, by comparing the difference in density between various kinds of matter to the differences between a dry sponge that is full of air and a wet sponge that is full of water. Both are full of matter, yet one is denser than the other. Even his contemporaries recognized the unsatisfactory nature of this explanation; and explaining differences in density remained a continuing problem for philosophers who denied the existence of void.

In the *Principia philosophiae* (*Principles of Philosophy*) (1644), Descartes gave a more detailed account of motion than Gassendi did in the *Syntagma philosophicum*. Descartes stated laws of motion that he thought would provide the foundations for a mathematical science of motion. Although the term "laws of nature" had been used since classical times, Descartes was the first person to identify particular propositions as laws of nature. He attempted to deduce the laws of motion from first principles, appealing to God's immutability. From the laws of motion, he attempted to derive mathematical laws of impact that describe how two bodies move after colliding with one another. Although even his contemporaries recognized the inadequacy of his laws of impact, the prominent place these laws hold in his system reflects their importance in a mechanical philosophy, according to which contact and impact are the only possible causes in the physical world. Having established the physics that he considered fundamental to his system, Descartes proceeded to give mechanical explanations of all the phenomena in the world, including the motions of the heavens, light, the qualities of material things, and even the human body.

Descartes was fully aware of the developments in astronomy and attempted to create a new cosmology in mechanical terms. Wanting to avoid Galileo's fate, he attempted to hide his Copernicanism. He actually suppressed his first book, *Le monde* (*The World*), written during the early 1630s, when he learned of Galileo's condemnation in 1633. When he finally published the *Principia philosophiae* in 1644, he did everything he could to disguise his real views. Remember that he denied the existence of the void. Because all bodies in his universe move in curved paths, he reasoned that their combined motions would produce huge whirlpools—he called them vortices—in the spaces between the heavenly bodies. Each star and the sun are at the center of a vortex. These bodies shine by their own light because the circular motions of the vortices produce a centrifugal tendency (a tendency to move out from the center) in the particles of subtle matter, and those moving particles cause us to perceive light. Because the outward motion of the smaller particles would leave empty spaces close to the center and because no empty spaces exist in Descartes' world, the particles further out in the vortex push the larger particles into the spaces closer to the center of the vortex, resulting in their weight or gravity.

The planets travel around the sun in the matter forming its vortex. How did this picture avoid the conclusion that Descartes endorsed Copernican astronomy? Descartes defined a body's "place" as the matter immediately surrounding it. Strictly speaking, the earth is not moving because it does not move with respect to the matter in the vortex that contains it. Descartes thus claimed that he avoided making the condemned assertion of the earth's motion.

Descartes hoped that the Jesuits would adopt his *Principia philosophiae* as a physics textbook in place of the Aristotelian texts they still used in their colleges. Posthumous condemnation of the *Principia philosophiae* in 1662 and its consignment to the Index of Forbidden Books by the Roman Catholic Church in 1663 in response to his attempt to give a mechanical explanation of the real presence of Christ in the Eucharist doomed that hope (posthumously) to disappointment.

Another natural philosopher, Thomas Hobbes (1588–1679), was the specter haunting the mechanical philosophy. His system reinforced fears that the mechanical philosophy would lead to materialism and atheism. In *The Elements of Philosophy* (1655), Hobbes propounded a complete philosophy of nature—including matter, man, and the state—based entirely on mechanistic principles. Although the details of his mechanical philosophy were not very influential, his mechanical account of the human soul and his thoroughly deterministic account of the natural world alarmed the more orthodox thinkers of his day.

## Explaining the World: Manifest and Occult Qualities

Having established—to their own satisfaction—that matter and motion suffice as the ultimate terms of explanation for natural phenomena, the mechanical philosophers tried to demonstrate that these terms could explain all the qualities of bodies. In this sense, the mechanical philosophy functioned as a language: it could provide explanations of any and all phenomena—real or not—that could possibly exist. The systematic treatises on the mechanical philosophy, such as Gassendi's *Syntagma philosophicum* or Descartes' *Principia philosophiae*, all contain sections in which they try to show how the various qualities of things could be explained in mechanical terms.

In adopting this approach to qualities, the mechanical philosophers were taking direct aim at the Aristotelian notions of real qualities and substantial forms. According to the traditional view, qualities—such as color, taste, and odor—really inhere in bodies. Thus, for example, a red body is really red; and no matter how finely one divided it, its matter would still be red. Mechanical philosophers claimed instead that the color of an object results from the effect the (uncolored) particles of light have on our eyes. Consequently, redness is never really in the observed body: the configuration of the particles of matter on its surface affects the motions of the material rays of light in such a way that their motions and configuration induce a sensation of red when they strike our eyes.

Medieval Aristotelians introduced the notion of substantial form to explain how compound substances can have qualities lacking in their individual components. For example, the components of gunpowder—sulfur and niter—do not possess the explosive power of the combined product. To explain how gunpowder and other compound substances (called "mixt bodies" in the seventeenth century) possess such additional properties, Aristotelian philosophers introduced the concept of the substantial form, a form giving the combination its special properties. Mechanical philosophers eliminated real qualities and substantial forms from their systems by explaining all qualities in terms of the motions and configurations of microscopic particles of matter and their effect on our sense organs.

Given the mechanical account of the world, the doctrine of primary and secondary qualities became the centerpiece of physics as well as of the theory of perception. According to the doctrine of primary and secondary qualities, matter actually possesses only a few qualities, the primary qualities. All other qualities—color, texture, odor, etc.—result from the interaction between configurations of the matter composing bodies and our senses. For Gassendi, the primary qualities are magnitude, figure, and heaviness; for Descartes, matter possesses only one primary quality, extension. The fundamental problem for mechanical philosophers was to explain the ways in which the corpuscular structures of physical things produce their secondary qualities and to understand our perception of those qualities.

For Gassendi, all bodies are alike insofar as they consist of atoms and void. An individual body possesses its particular qualities because of the arrangement of its constituent atoms. Gassendi reinterpreted the Aristotelian concepts of substance and accident, matter and form, in atomistic terms: for him, the form is nothing but the configuration of the atoms composing the object. The configuration of the atoms and the void of which the object consists explains all its other qualities. Qualities become manifest in two ways: by the effects they produce on our organs of sense and by the effects they produce on other material bodies and thus by the changes they produce in our sensations of these bodies.

Gassendi tried to explain all kinds of qualities mechanically: rarity and density, transparency and opacity; magnitude, figure, subtlety, smoothness, and roughness; mobility; gravity and levity; heat and cold; fluidity and hardness; moistness and dryness; softness, rigidity, flexibility, elasticity, and ductility; taste and odor; sound, light, and colors. For example, he explained the phenomena of light by means of a particulate model. Comparing a ray of light to a stream of material particles, he noted that just as a bean or another material particle rebounds from a wall because it cannot move through space already occupied by another body, so a material ray of light is reflected from material bodies by impact. The fact that some bodies are transparent and allow light to pass through them might seem like a counterexample to the corporeality of light. But Gassendi explained this phenomenon in terms of his atomism by arguing that a body is transparent because there are many pores separating its constituent atoms; thus, the particles of light can pass easily between them.

Consider the analogy of a sieve. Drop a handful of sand from some distance onto a sieve; those grains that fall on the holes go straight through; those that fall on the solid parts rebound. So too when a ray of light falls on some solid body: those corpuscles that fall on the solid parts bounce back, and those that fall on the pores pass through the body. Sometimes a pore does not penetrate all the way through the body but may twist around in its interior. Particles of light that enter such pores are lost in the interior of the body, thus accounting for the absorption of light. Using a similarly rich mix of observation, analogy, speculation, and fantasy, Gassendi tried to show how every kind of known phenomenon could be explained solely in terms of matter and motion.

In their determination to encompass all phenomena within the mechanical categories and to banish action-at-a-distance from the natural world, the mechanical philosophers sought to explain the traditionally occult qualities in mechanical terms. Unlike modern usage, according to which the word "occult" refers to something magical or spiritual, in the early modern period the word simply meant "hidden." So-called occult qualities had hidden, not immediately evident causes. Gassendi argued that mechanical principles could explain the occult qualities just as they can explain the so-called manifest qualities. Our ignorance of the mechanisms that produce them does not justify the conclusion that the normal processes of nature do not cause them.

Examples of such occult qualities include all of the so-called sympathies and antipathies in the world, many of which Gassendi proceeded to explain in mechanical terms. The occult sympathies that had been invoked to account for otherwise inexplicable attractions—such as the fact that rubbing amber causes it to attract bits of straw—received mechanical explanations: the attraction is caused by "hooks, cords, goads, prods, and other such things, which although they are invisible, must not be called nothing." For example:

> When you observe a chameleon seize a fly from half a palm away and draw it to its mouth, you see an organ of attraction, the vibration and retraction of the tongue by its great agility, the end of which is viscous and curves into itself. What would you otherwise judge to happen when, for example, amber, sealing wax, and other electrics, when you first rub them, seize, draw, and hold straw and other

light things? Indeed, innumerable little rays like tongues seem to be emitted from electric bodies of this kind, which they fill, seize, and carry back and hold, by the insinuation of their ends into the little pores of those light things.[1]

All sorts of remarkable phenomena may incite wonder in us: the heliotrope's following the sun, the reported impossibility of tuning a musical string made of sheep gut in perfect consonance with one of wolf gut, the poisonous glance of the basilisk, the charming of snakes with music, the electric glow of the torpedo fish, the strange power of the remora to bring a ship to a dead stop, the medicinal properties of various substances, and the long-distance healing power of the weapon salve. Yet there is no reason to believe that these unusual phenomena have any cause other than what, according to Gassendi, produce the most familiar effects, namely, the motions and collisions of atoms in the void.

Descartes also enumerated all the qualities of bodies, whether manifest or occult, and provided mechanical explanations for them. Like Gassendi and other mechanical philosophers, Descartes showed how these qualities— gravity, light, heat, the rarefaction and compression of air, the freezing and evaporation of water, the tides, the particular properties of mercury and other chemical substances and metals, the nature of fire and of gunpowder, the properties of glass, of the magnet, the amber effect, and finally sensation—could be explained in terms of the motions and configurations of the particles of matter composing bodies.

Descartes pointed especially to his putative success in explaining magnetism, traditionally a paradigm case of an occult quality. Descartes' knowledge of the properties of magnets came from the empirical studies of William Gilbert (1544–1603), who had described the results of his experiments in *De magnete* (*On the Magnet*) (1600). According to Gilbert, magnets exhibit polarity, possessing north and south poles, whereby likes repel and unlikes attract. Among metals and other substances, only iron can be magnetized. Magnets, such as the needle of the compass, orient themselves in relation to the earth. And iron filings scattered around a magnet will move into a characteristic pattern surrounding the magnet. Earlier natural philosophers had explained these phenomena by endowing iron and magnets with innate activity of one kind or another.

Mechanical philosopher that he was, Descartes proposed a mechanistic

1. Pierre Gassendus, *Syntagma philosophicum*, in Petrus Gassendi, *Opera omnia*, 6 vols. (1658; facsimile repr., Stuttgart–Bad Cannstatt: Friedrich Frommann Verlag, 1964), 1:450.

explanation. He suggested that the earth emits screw-shaped particles from each of its poles. From one pole it emits right-handed screw-shaped particles, and from the other left-handed ones. Correspondingly shaped pores in the earth cause these particles to flow in regular patterns. A piece of iron that has remained buried in the earth for a period of time develops screw-shaped pores, as the particles move through it. It is thus magnetized and will become oriented by the flow of these magnetic particles as they move around the earth.

Having given mechanical explanations of both the manifest and occult qualities, Descartes claimed that he had thus explained all the phenomena of nature.

> And thus, by simple enumeration, it is concluded that no phenomena of nature have been omitted by me in this treatise. For nothing is to be counted among the phenomena of nature, except what is perceived by sense. However, except for size, figure, and motion which I have explained as they are in each body, nothing located outside us is observed except light, color, odor, taste, sound, and tactile qualities; which I have now demonstrated are nothing in the objects other than, or at least are perceived by us as nothing other than, certain dispositions of size, figure, and motion.[2]

Although Descartes' cosmology—and for that matter his entire philosophy of nature—may seem fantastic today, his contemporaries took it very seriously. His mechanical philosophy was the first fully articulated account of the universe since Aristotle's, and it provided seemingly satisfactory answers to many of the most pressing questions of the day.

Gassendi's and Descartes' programmatic works on the mechanical philosophy set the agenda for the next generation of natural philosophers, who accepted mechanical principles in general and believed that they had to choose between Gassendi's atomism and Descartes' plenism. Some of the most prominent natural philosophers of the second half of the seventeenth century—Robert Boyle (1627–91), who invented the term "mechanical philosophy"; Christiaan Huygens (1629–95); and Isaac Newton—developed their philosophies of nature in this context.

In presuming that all natural phenomena can be explained in terms of matter and motion alone and in ruling out action-at-a-distance, the mechanical philosophers departed from traditional philosophies of nature that had endowed matter with various kinds of activity. Aristotelians had thought that

---

2. René Descartes, *Principles of Philosophy*, trans. Valentine Rodger Miller and Reese P. Miller (Dordrecht: Reidel, 1983), pp. 282–83.

there exist natures which give bodies tendencies to move in characteristic ways. For example, according to Aristotle heavy bodies, because of their nature, tend to move toward their natural place at the center of the world; likewise, the form of the oak tree, potentially contained in the acorn, causes the constituent matter to grow into an oak tree, rather than a maple or a eucalyptus. Many Renaissance philosophers in the Neoplatonic, Hermetic, and Paracelsian traditions portrayed a highly animistic world, characterized by sympathies and antipathies, which act at a distance and endow the material world with its own innate activity. The mechanical philosophers rejected the activity of matter because they thought that active matter, insofar as it is self-moving, seemed capable of explaining phenomena without any reference to God. They thought that they could avoid this danger of atheism because naturally inert matter requires an external source of motion, a motive source that lies outside the natural, material realm. God, as the only source of motion, seemed absolutely necessary to most seventeenth-century mechanical philosophers.

## Including the Divine: Theological Dimensions

While many thinkers found the mechanical philosophy attractive, they found some of its possible consequences theologically troubling. Some natural philosophers feared that the mechanical philosophy would lead to materialism (the belief that only matter exists) or deism (the belief that God created the world and the laws of nature but has since withdrawn from any active role) by denying the doctrines of creation and divine providence. The ancient association between Epicurean atomism and atheism reinforced these fears. Because of the medieval synthesis of Christian theology and Aristotelian philosophy, some theologians and philosophers perceived the rejection of Aristotelianism as a challenge to fundamental tenets of Judeo-Christian religion. Christian mechanical philosophers adopted a variety of strategies to defend themselves against these perceived threats, including frequent appeal to the argument from design as a way of establishing God's providential relationship to the creation, special attention to proving the existence of an immaterial, immortal human soul, and attempts to give a mechanical explanation of the real presence of Christ's body and blood in the elements of the Eucharist.

The philosophy of Hobbes, whose materialism and determinism became emblematic of the theological dangers posed by the mechanical philosophy, made the anti-providential consequences of the mechanical philosophy real. Fear of "Hobbism" led other thinkers to insist on providential interpretations of the mechanical philosophy. Gassendi, who (as we have seen) modified atomism to rid it of the materialistic and atheistic associations with Epicureanism, explicitly incorporated divine providence into his version of the mechanical philosophy, partially as a response to Hobbes, whom he had met in Paris in the 1640s. Making extensive appeal to the argument from design, Gassendi reasoned that the world must be the product of intelligent design rather than the chance collision of atoms, as both the ancient atomists and Hobbes believed. Denying both the Epicurean doctrine of chance or fortune and the Stoic doctrine of fate, Gassendi redefined these notions providentially, interpreting fortune as an expression of divine foresight and providence, and fate as divine decree.

Ensuring a role for a supreme being in the world was one problem. Interpreting God's relationship to the creation was another, one that had major implications for understanding the metaphysical and epistemological status of the laws of nature and miracles. Is God bound by his creation, or is he always free to change whatever he created in the world? The seventeenth-century answers to this question grew out of thirteenth- and fourteenth-century discussions following the introduction of Aristotle's philosophy into mainstream European thought. A delicate balance existed in medieval theology between the rationality of God's intellect and his absolute freedom to exercise his power and will. Intellectualist theologians like Thomas Aquinas, who emphasized God's rationality, were more inclined to accept elements of necessity in the creation than voluntarists like William of Ockham (ca. 1285–1349), who emphasized God's absolute freedom and concluded that the world is utterly contingent on divine will.

Seventeenth-century thinkers transformed these ideas about God's relationship to the creation into views about the metaphysical and epistemological status of human knowledge and the laws of nature. For intellectualists the laws of nature describe the essences of things and we can know them *a priori* (prior to experience); the empiricist and probabilist interpretations of scientific knowledge adopted by the voluntarists provided a way of thinking about the contingency of a world that no longer contains essences. For the voluntarists, the laws of nature are simply descriptions of regularities observed in nature. According to them, God's power to intervene in the creation remains unrestricted, implying that the laws of nature are contingent truths that always remain subject to alteration. God's continuing freedom to intervene in the natural world implies that no necessary connections exist that would underwrite the possibility of a priori knowledge. Consequently, the human capacity to know about the world must rely on observation and empirical methods.

Gassendi, who was a voluntarist, described a world that is utterly contingent on divine will. This contingency expressed itself in his conviction that empirical methods are the only way to acquire knowledge about the natural world and that the matter of which all physical things consist possesses some properties that can be known only empirically. The laws of nature simply describe empirical generalizations. God can change them at will, a fact to which miracles attest. Descartes, as an intellectualist, described a world in which God had embedded necessary relations, some of which enable us to have a priori knowledge of substantial parts of the natural world. The capacity for a priori knowledge extends to the nature of matter, which, Descartes claimed to demonstrate, possesses only geometrical properties. According to Descartes, the laws of nature are necessary truths that follow directly from God's attributes, specifically his immutability.

The Cambridge Platonists, like Henry More (1614–87), adopted an even more extreme form of intellectualism, according to which absolute standards of goodness and mathematical relationships that exist independently of God limit divine freedom. More initially found Cartesianism particularly attractive because Descartes considered spirit to be as real as matter. He eventually grew very critical of Cartesianism, fearing that its sharp separation of matter from spirit might, in the end, lead to materialism. More argued that all sorts of phenomena are impossible to explain simply in terms of "the jumbling together of the *Matter*."[3] Such phenomena include the parallelism of the axis of the earth and the consequent sequence of the seasons, gravity, the behavior of air in Boyle's experiments with the air pump, and all the evidence of design in the parts and habits of living things. Resisting purely mechanical explanations in favor of the actions of a wise providential God, More explained such phenomena by a Spirit of Nature which is incorporeal, extended, and indivisible, a causal agent carrying out God's providential plan for the creation.

More set out to prove the reality of spirits, to defeat what he perceived as Hobbes' materialism as well as what he thought were the atheistic consequences of Descartes' philosophy. In addition to theological and philosophical arguments, More sought empirical evidence that spiritual entities could cause changes in material bodies. He reported many accounts of apparitions and witchcraft, which he thought lent credence to the existence of immaterial spirits. Thinking along the same lines, Joseph Glanvill (1636–80) urged the newly founded Royal Society (one of the earliest scientific societies) to under-

3. Henry More, *An Antidote against Atheism; or, An Appeal to the Natural Faculties of the Mind of Man, Whether There Be Not a God*, 3rd ed. (London, 1662), p. 47.

take research projects to study haunted houses and alleged cases of witchcraft in order to find empirical evidence for the reality of incorporeal spirits and provide a compelling case against materialism.

Robert Boyle, a deeply religious man, discussed the theological implications of his corpuscularianism at great length. He believed that God had created inert matter and had endowed it with motion. God created laws of nature but could violate those laws at will; biblical miracles provided evidence for that claim. Boyle believed that, far from leading to atheism and materialism, natural philosophy enhances one's appreciation of God's wisdom, power, and goodness. Citing Plato, Boyle declared, "The World is Gods Epistle, written to Mankinde."[4] He thought that the study of nature as an act of worship leads to greater knowledge of the Creator by directly acquainting the careful observer with God's wisdom and benevolence in designing the world. The astute observer can discern God's purposes everywhere. God is not entirely knowable, however. Boyle carefully acknowledged the limits of human reason in theology and argued that those limits extend to natural philosophy as well. Human knowledge could attain neither certainty nor completeness.

Boyle's idealized "Christian Virtuoso" discovers the deep connections between natural philosophy and Christian theology. He even considered how natural philosophy would be completed in the afterlife, until the time of the final judgment and the general resurrection, when God would create "a new heaven and a new earth," possibly changing all the laws of nature that hold in the present, pre-millennial world. His insistence on divine freedom underpinned his conviction that nothing in the natural world limits God's freedom. And he regarded the mechanical philosophy as more compatible with Christianity than Aristotelianism or other worldviews.

Like other mechanical philosophers, Boyle wanted to avoid the dangers of materialism, and so he emphasized the boundaries of mechanization. He believed that the world contains spiritual entities that cannot be explained in terms of matter and motion. He claimed that human souls are entirely incorporeal. Thus, each human life involves a miraculous intervention by God and incorporates a nonmaterial entity at its core. The philosophers' stone (the alchemical agent of transmutation) also served to define the limits of mechanization for Boyle. Not only could it bring about transmutations, but it could

4. Robert Boyle, *Some Considerations Touching the Usefulnesse of Experimental Naturall Philosophy* (1663), in *The Works of Robert Boyle*, ed. Michael Hunter and Edward B. Davis, 14 vols. (London: Pickering & Chatto, 1999–2000), 3:233.

also attract angels and other spirits. He believed that this power of the philosophers' stone provided a strong argument against atheism.

Boyle sought to promulgate his belief in the close relationship between science and religion. His will provided funds to establish the Boyle Lectures, a series of eight lectures a year, the purpose of which was to confute atheism. Several prominent scholars and divines were among the Boyle lecturers, including Richard Bentley (1662–1742), Samuel Clarke (1675–1729), William Whiston (1667–1752), and William Derham (1657–1735).

Despite the controversies within and surrounding the mechanical philosophy, it provided a new conceptual framework for the developments in many aspects of seventeenth-century natural philosophy. It established important parts of the agenda for studies of motion, light, and the nature of matter in the generations to follow.

# 5 Shifting Boundaries

*From Mixed Mathematics to Mathematical Physics*

Both the science of motion and the science of vision held marginal positions in the Aristotelian classification of the sciences as branches of mixed mathematics. According to that scheme physics (natural philosophy) explains changeable, sensible things by reference to their causes. Mathematics, which deals with changeable, insensible things, proceeds by deductions from axioms that are known to be true prior to any empirical experience. Mechanics, which at the time dealt with the mathematical analysis of simple machines (the inclined plane, pulley, wedge, lever, and screw), and the science of vision, which dealt with the geometry of light rays and the physiological process of vision, both used mathematics to describe empirical (sensible) facts about physical (changeable) phenomena. Consequently, mechanics and the science of vision carried the designation "mixed mathematics."

During the early modern period, both of these subjects underwent major transformations. The abstract consideration of motion became linked with the study of simple machines, resulting in a new concept of motion that made the mathematical study of physical motions possible. At the same time, the study of light became separated from the study of vision. Natural philosophers came to use mathematics to describe empirical phenomena without seeking causal explanations in the Aristotelian sense. This development altered traditional disciplinary boundaries.

## Moving Bodies: The Science of Motion

The concept of motion plays a fundamental role in virtually all philosophies of nature. Three problems about motion dominated the writings of natural philosophers and mathematicians in the early modern period: (1) formulating what came to be called the principle of inertia; (2) describing the impact of bodies; and (3) accounting for circular motion. The Copernicans' claim that heliocentric astronomical theory describes physical reality demanded both a response to the traditional objections to the earth's motion and a physically adequate account of circular motion. Furthermore, because mechanical philosophers reduced causality to the contact and impact of material bodies, they recognized that the problem of impact was central for physics. The struggle to solve these problems involved redefining basic terms such as "gravity," "mass," and "force"—terms that had a long history in the analysis of motion.

The foundations of a new science of motion emerged from the work of both Galileo and Descartes. Galileo addressed particular problems—notably the speed of falling bodies and the trajectory of projectiles—while Descartes explicitly attempted to formulate a new philosophy of nature based on a new concept of motion. Together, their ideas led to the breakdown of the Aristotelian distinction between natural and violent motion, a development that facilitated the application of mechanical ideas to the explanation of the natural world.

According to traditional Aristotelian thinking, mechanics was a practical science that made use of motions that are contrary to nature. Indeed, the Greek word *mēchanē* means "trick." These motions can trick nature into doing things outside of its usual course. The Aristotelian text *Quaestiones mechanicae (Mechanical Problems)*, most likely not written by Aristotle himself, opens with the following distinction: "Remarkable things occur in accordance with nature, the cause of which is unknown, and others occur contrary to nature, which are produced by skill for the benefit of mankind," as, for example, when a pulley uses downward motion to cause a heavy body to rise.[1] Although this book was probably unknown during the Middle Ages, after the humanist Fausto Vittore (1480–1551?) translated it into Latin in 1517 many commentators discussed its explanations of simple mechanical devices. Among the discussions that the *Quaestiones mechanicae* spawned, the disciplinary status of mechanics occupied a central place. By the end of the sixteenth century, some influential commentators began to blur and finally to eliminate the traditional distinction between nature and machines, that is, between the subject matter of physics and that of mechanics. Both Galileo and Descartes read these commentaries on the *Quaestiones mechanicae*. Thus, Renaissance commentaries on a pseudo-Aristotelian text directly influenced the development of a new science of motion, the emergence of the mechanical philosophy, and the resulting reclassification of the sciences in the seventeenth century.

Galileo's physics was the crucial link between the developments in Copernican astronomy and the articulation of a new philosophy of nature. His new science of motion provided answers to the objections to the earth's motion,

1. Aristotle, *Mechanical Problems,* in *Minor Works,* trans. W. S. Hett (Cambridge: Harvard University Press, 1936), p. 331.

and it provided the foundation for the new concepts of cause and change embodied in the mechanical philosophy. Galileo's approach to motion grew directly out of his education. During his undergraduate years, contact with the mathematician Ostilio Ricci (1540–1603) introduced Galileo to the rigorous mathematics of Euclid (fl. ca. 300 BC) and the ideas of the Greek mathematician Archimedes (287?–212 BC). Ricci was a student of Niccolò Tartaglia (1500–1557), who had edited and published a Latin translation of the works of Archimedes in 1543. Archimedes' methods deeply influenced Galileo's thinking. Although Galileo rarely cited any other thinkers, ancient or modern, he frequently mentioned "the divine Archimedes."

Galileo found three aspects of Archimedes' work particularly important: the science of weight, or statics; his method for calculating areas of plane figures; and his study of the behavior of floating bodies, or hydrostatics. Archimedes had worked out the law of the lever geometrically, having developed a method for determining centers of gravity. The law of the lever gave him a conceptual tool that he used to solve a variety of problems in mechanics. For example, Archimedes developed methods for calculating the areas of curved figures by dividing them into polygons with more and more sides and then summing up the total areas of the parts of the polygon, a technique known as the method of exhaustion because, as the number of sides of the polygon increases, the difference between the area of the curved figure and the area of the polygon is exhausted. In order to check the accuracy of his proofs, he constructed material models of the figures in question and placed them on a balance to determine the equivalence of the area of the many-sided polygon with that of the curved figure. In his treatise *On Floating Bodies*, he used the principles of statics to calculate whether bodies in fluids will sink or float.

In 1590, while teaching at the University of Pisa, Galileo started writing a treatise, *De motu* (*On Motion*), in which he attempted to apply Archimedes' mathematical methods to the traditional account of motion. Galileo began *De motu* with the Aristotelian assumption that all motion must have a cause. He identified that cause as the natural heaviness possessed by all bodies. This assumption represented a departure from Aristotelian physics, which held that some bodies are naturally heavy and some are naturally light. By endowing all bodies with the same property, heaviness, Galileo made it possible to apply mathematics to the cause of motion, because all bodies possess the same property in different quantities. In *De motu*, Galileo retained the Aristotelian distinction between natural and violent motion. The body's heaviness causes it to move naturally downward, toward the center of the earth. Violent, upward motion has an external cause. If one thinks of bodies as moving through a medium, then some bodies move upwards because the heavier bodies push them up and out of the way as the heavier bodies move down. Galileo used specific weight (the weight of the body divided by its volume) as the quantity that determines the relative motions of bodies. He used the model of the balance or lever, something he knew from his study of Archimedes, to calculate the motions of bodies in resisting media. Galileo thought that the speed of a body's natural motion is a function of its specific weight minus the resistance of the medium, or $V = W - R$. If $R = 0$ (that is, if the motion occurs in a void), the body's weight will fully determine its natural motion. Galileo thus asserted the possibility of motion in a void and identified it as the body's characteristic motion.

Using this formulation, Galileo considered various problems about motion. He discussed the nature of free fall, the possibility of motion in a void, and the perennially recalcitrant problem of projectile motion. In the end, however, he realized that certain motions are neither natural nor violent. For example, a ball rolling on the spherical surface of the earth is moving neither toward nor away from the center of the earth. Because it is not seeking a natural terminus (the center of the earth), and because it is not resisting its natural tendency to fall down, its motion could, in principle, continue indefinitely. This example and others like it probably caused Galileo to stop writing *De motu* and rethink the underlying concept of motion. He left the treatise unfinished and unpublished.

By 1604, Galileo had come up with an entirely new understanding of motion. He worked out his new theory over many years, finally publishing it in his *Dialogue Concerning the Two Chief World Systems* (1632) and *Discourses and Mathematical Demonstrations Concerning Two New Sciences* (1638). Galileo's new science of motion was kinematics, the study of motion in terms of time and distance without considering its cause. This approach departed totally from Aristotelian physics and the physics of *De motu*, in which analysis had focused on the cause of motion—the heaviness, lightness, or specific weight of the moving body.

Galileo gave rigorous definitions of relevant quantities, such as uniform velocity and uniform acceleration. He demonstrated that the motion of a body falling with uniform acceleration can be described as

$$s \propto t^2$$

where $s$ (*spatium* in Latin) represents the distance traversed and $t$ represents the time elapsed during the fall. Galileo proved this law of free fall mathematically. By analyzing the behavior of bodies rolling down inclined planes, he established that naturally falling bodies are uniformly accelerated. When asked about the cause of this acceleration, he vigorously maintained his kinematic approach: "The present does not seem to me to be an opportune time to enter into the investigation of the cause of the acceleration of natural motion, concerning which various philosophers have produced various opinions." "For the present," he continued, it was sufficient to "investigate and demonstrate some attributes [*passiones*] of a motion so accelerated (whatever be the cause of its acceleration)."[2]

Galileo made another extremely important assumption in his analysis of motion, a first attempt to formulate what would become the principle of inertia. Already in *De motu*, he had realized that a body moving with uniform velocity on the surface of the earth would be neither natural nor violent. Therefore, neither resisting its natural downward tendency nor coming to an end at the center of the earth, it would continue moving forever unless something (for example, friction) interfered with its motion. Galileo put this principle to good use. In his mature science of motion, he combined it with his law of free fall to prove that the trajectory of a projectile is parabolic.

This result is particularly important because it enabled Galileo to answer some of the most compelling objections to the earth's motion. Consider the argument that if the earth rotated on its axis, a heavy body dropped from the top of a tower would not land at the base of the tower, but would be left behind. This objection to the earth's motion was based on the Aristotelian assumption that a body cannot undergo two different motions at the same time. Galileo countered this objection by noting that when the body is dropped from the top of the tower, it continues its motion "parallel" to the earth's surface because it does not encounter any resistance to that motion. That is to say, it continues to share the motion of the earth that it had when it was held at the top of the tower. Consequently, it will keep up with the motion of the earth even after it is dropped from the tower. The body will indeed fall to the base of the tower even if the earth is moving.

This reasoning introduced a new metaphysics of motion. In Aristotelian physics, the motion of a body depends on the body's form or nature. Heavy bodies fall down; light ones rise. In Galilean physics, all bodies fall according to the same law. Consequently a body's motion reveals nothing about its nature. Nevertheless, Galileo's break with Aristotle was not complete. He retained a

2. Galileo Galilei, *Two New Sciences, Including Centers of Gravity and Force of Percussion*, trans. Stillman Drake (Madison: University of Wisconsin Press, 1974), pp. 158–59.

## Galileo's Demonstration of the Parabolic Trajectory of Projectile Motion

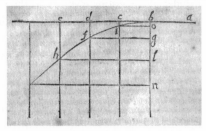

Consider a body moving from right to left on the horizontal plane $ab$ at uniform speed. The plane ends at $b$, and the body goes into free fall. But, at the same time that it falls, it loses none of its horizontal motion. Therefore, at equal intervals of time, corresponding to its passing through the equal intervals of distance $bc$, $cd$, and $de$, the body will fall according to the law of free fall, namely, $s \propto t^2$. To state this in algebraic terms, which Galileo did not use, the trajectory can be described as $x =$ $-y^2$, and that is the equation describing a parabola.

■ Reprinted from Galileo Galilei, *Discorsi e dimonstrationi mathematiche, intorno à due nuouo scienze, attenenti alle mechanica, & I mouimenti locali* (Bologna: de Dozza, 1655), p. 184

concept of natural motion—the downward motion of heavy bodies—a concept in which an idea of natural place was still implicit.

Gassendi developed this new conception of motion further by clearly stating that rectilinear, rather than "horizontal" (parallel to the earth's surface), motion would continue forever unless impeded by something else. He considered a body moving in void space. "If anyone would push it, it would move in whatever direction the push was made, uniformly and according to the slowness or speed of the impulse, and indeed perpetually in the same line, since there is no cause which would accelerate or retard it unless by diverting its motions." He continued: "Whence also we conclude that every motion once impressed is by its nature indelible, and does not diminish or cease unless through an external cause which checks it."[3] Thus Gassendi eliminated the concept of natural motion and the existence of the center as a qualitatively privileged location in space.

The new concept of motion embodied a fundamental rejection of Aristotelian physics and cosmology. The cosmos no longer had a center that determines the nature of motions. Consequently, there were no longer any

3. Gassendi, *Syntagma philosophicum*, trans. and quoted in Richard S. Westfall, *Force in Newton's Physics: The Science of Dynamics in the Seventeenth Century* (New York: American Elsevier, 1971), pp. 101, 102.

privileged directions of motion, the up and down of Aristotelian physics. In contrast to Aristotelian physics, according to which all motions require a causal explanation but a body at rest in its natural place does not, motion and rest now became simply two states of the same kind of change. Inertial motion and inertial rest require no explanation; only a change in a body's state of motion or rest requires a cause. Bodies are indifferent to motion or rest. The causes of all changes in their state of motion or rest are external to them. This new understanding of motion eliminated the ordered Aristotelian cosmos. With no center and no privileged directions, space is the same in all directions. Motion and space can both be described mathematically.

Descartes' laws of nature codified the new science of motion. He proposed a principle of the conservation of motion as the foundation of his analysis of motion: "God is the primary cause of motion; and . . . He always maintains an equal quantity of it in the universe."[4] This principle described the quantity of motion as the size of the body multiplied by its speed ($mv$). Given this foundation, Descartes stated the laws of motion as follows:

> The first law of nature: that each thing, as far as is in its power, always remains in the same state; and that consequently, when it is once moved, it always continues to move.

> The second law of nature: that all movement is, of itself, along straight lines; and consequently, bodies which are moving in a circle always tend to move away from the center of the circle which they are describing.

> The third law: that a body, upon coming in contact with a stronger one, loses none of its motion; but that, upon coming in contact with a weaker one, loses as much as it transfers to that weaker body.[5]

His first two laws jointly articulated a principle of inertia that was free from lingering Aristotelian assumptions about natural motion and motion around a center. He clearly stated that what we call inertial motion is rectilinear.

Following the third law of nature, Descartes stated seven rules governing different cases of impact. The first rule dealt with the case of two equal bodies moving towards each other with equal speeds. Following impact, their speeds would remain the same, but the direction of their motions would be reversed. The subsequent rules dealt with cases of impact between bodies of different

    4. René Descartes, *Principles of Philosophy*, trans. Valentine Rodger Miller and Reese P. Miller, (Dordrecht: Reidel, 1983), p. 57.
    5. Ibid., pp. 59–61.

sizes approaching each other at different speeds. Descartes' attempt to mathematize impact did not succeed—as Huygens and other seventeenth-century natural philosophers amply demonstrated—but his work stimulated others to find better solutions to the problem.

Christiaan Huygens proved the inadequacy of Descartes' rules of impact. He followed Descartes in attempting to apply mathematics to physics and the mechanical philosophy and in ascribing only geometrical properties to matter, but he differed from Descartes in accepting the existence of void space, in which the small particles of matter move. He addressed several of the most pressing problems in the seventeenth-century science of motion, including developing a mathematical analysis of circular motion and an improved theory of impact. Huygens consistently applied the concept of the relativity of motion—which Descartes had stated but had not applied—to the problem of impact. He also used the principle, first developed by Evangelista Torricelli, that the center of gravity of two colliding bodies always moves uniformly in a straight line.

Reasoning from these principles, Huygens demonstrated the inconsistency of Descartes' account of impact. Relativity, in the case of impact, requires that one take account of the direction of motions. Descartes' principle of the conservation of motion assumed an absolute frame of reference, an assumption incompatible with the relativity of motion. Having demonstrated the errors of Descartes' rules of impact, Huygens derived a different formulation, one which eliminated his predecessor's concept of the conservation of motion: "When two bodies strike each other, the quantity which is obtained by adding the products of the magnitudes of the individual bodies multiplied by the squares of their velocities, is found to be the same before and after the impact of the bodies."[6] The problem of impact remained central to the investigations of natural philosophers and mathematicians during succeeding generations. Lurking behind Descartes' and Huygens' kinematic approach to the problem was the question of what actually changes the direction and speed of the bodies in impact. The need for a concept of force essentially undermined the attempt to deal with motion in strictly kinematic terms.

Circular motion presented a second problem of critical importance to seventeenth-century developments in both astronomy and natural philosophy. With the claim that astronomical theories actually describe the physical world came the need for a physical account of orbital motion. A first step

6. Christiaan Huygens, *De motu ex percussione*, trans. and quoted in Westfall, *Force in Newton's Physics*, p. 157.

toward addressing this problem was to figure out a mathematical description of circular motion. Descartes tried to solve this problem by using his laws of motion. Although his second law of motion states that bodies tend to move in straight lines, bodies in the universe described in *The Principles of Philosophy* actually move in closed curves because they are moving in a plenum. Since there is no empty space in Descartes' world, the matter surrounding bodies always obstructs a body's motion. In order for motion to occur, the obstructing bodies must move out of the way, and the bodies obstructing the motion of these bodies must move accordingly. The net result is that all motion is in closed curves. Descartes tried to describe the nature of circular motion. He thought that bodies moving in circular motion, like a stone on the end of a rope, have a tendency to move away from the center. Although he considered several examples to illustrate this point, he did not succeed in coming up with a mathematical description of this tendency.

Huygens successfully stated this relationship mathematically. He considered a circular motion, such as a stone on the end of a rope, and observed that at every moment the moving body has an inertial tendency to move in a straight line along the tangent to the circle. He called this tendency "centrifugal force"—the force pulling the body away from the center of the circle. By carefully considering the geometry of the situation, he derived a mathematical formula for centrifugal force:

$$F = \frac{m}{r} v^2,$$

where $m$ is the mass of the body, $v$ is its angular velocity, and $r$ is the radius of the circle in which it is moving.

Solutions to each of the problems about motion implicitly called for a concept of force. In the case of inertial motion, one could ask what external cause would make the body deviate from inertial motion or rest. When considering the impact of two bodies, one could ask what causes the colliding bodies to affect each other's motion. And, as becomes evident in Huygens' treatment of circular motion, one could ask how it is possible to explain the body's tendency to move in a tangent away from the circle.

The concept of force, however, raised a red flag for mechanical philosophers. They assumed that matter is inert, that it does not possess any activity of its own, and that there is no action-at-a-distance. Because they shared this understanding of matter, they had attempted to deal with motion kinematically—that is, in terms of distance and time without any consideration of the causes of motion. Galileo had stated this approach explicitly, and both

Descartes and Huygens attempted to continue solving problems about motion without considering causes. Despite their best intentions, unacknowledged concepts of force crept into their analyses. In the end, the development of a coherent and powerful science of motion required a concept of force, a radical modification of the mechanical philosophy.

## Illuminating the World: Optics and Theories of Vision

The study of optics arose from the question, How do I see? Ancient Greek natural philosophers proposed two possible physical explanations of vision: that something external enters the eye to produce vision (the intromission theory); or that something is emitted by the eye, strikes the object and renders it visible, and then something returns to the eye to cause vision (the emission theory). Both explanations raised the question of what it is that enters the eye. Aristotelians identified as the vehicle of vision the so-called visible species that transmit the object's form (in this case its visible form) to the eye. The atomists believed that an outer layer of the object's atoms, a *simulacrum*, traveled from the object into the eye. According to either theory, we see the object as a whole by immediate contact. Both Euclid (fl. 300 BC) and Ptolemy (ca. AD 100–170) thought that the eye emits some sort of rays that make things visible. This seemingly odd theory makes sense if we remember that we cannot see an object unless we actually look at it.

In the separate disciplinary tradition of mixed mathematics, the mathematicians Euclid, Hero of Alexandria (fl. AD 62), and Ptolemy approached the problem of vision geometrically. They assumed that light travels in straight lines and that they could analyze vision by tracing the passage of light from the visible object to the eye. Geometrical optics developed further in the hands of the Arabs. The most influential medieval writer, 'Abu 'Alī al-Hasan ibn al-Hasan ibn al-Haytham, known in the Latin West as Alhazen (the same Alhazen who had made such great contributions to astronomy), adopted an intromission theory—namely, the theory that vision occurs when light enters the eye. He combined the Greek practice of ray-tracing with his theory of vision by introducing a point-by-point analysis of object and image. Instead of considering the object and image as wholes as the Greeks had done, he considered each point on the object and traced the path of light from that point to the eye, thereby showing how the image is constructed point by point within the eye. Two thirteenth-century writers, the Franciscan John Pecham (ca. 1230–92) and the Polish Neoplatonist Erazmus Ciolek Witelo (fl. 1250–75), adopted Alhazen's ideas. Both men wrote treatises entitled *Perspectiva* (*On Optics*), in which they extended Alhazen's theory of vision.

Johannes Kepler appropriated these medieval ideas and developed them further, thus laying the foundations for seventeenth-century research in optics. He used Alhazen's point-by-point analysis, and he addressed the problem of finding a mathematical law of refraction in order to explain Galileo's telescope. In considering vision, Kepler compared the eye to a *camera obscura*, basically a pinhole camera, and used his point-by-point analysis to demonstrate that the image formed on the back of the eye is an inverted image of the object. Kepler's work was significant particularly because he applied mathematics to find solutions to physical problems.

Kepler's work on optics directly influenced Descartes, who, from his earliest writings on natural philosophy, addressed the problem of explaining light and vision in terms of matter and motion. Descartes' theory of matter lay at the heart of his explanation of light and vision. Because he endowed matter with only one intrinsic property, extension, the only way he could distinguish among different kinds of matter was by the size of their constituent particles. He thus described three kinds of matter, which he called elements. The first element, the element of fire, consists of the smallest particles, which move extremely swiftly and can take on any shape whatsoever. Because Descartes denied the existence of void, he thought that these smallest particles fill all the spaces and interstices between the larger particles that constituted the other elements. The second element, which he claimed to be the element of air, is also a very subtle fluid, consisting of spherical particles, joined together like grains of sand or dust. Although small compared to the particles of the third element, they are large compared to the first. The third element, the element of earth, consists of much larger particles, which hardly move in relation to each other. Luminous bodies, like the sun and the stars, consist of the first element, while the earth, comets, and planets, which do not emit their own light, consist largely of the third element. The second element is the medium through which light travels.

Descartes' new physics provided the necessary explanatory framework for his analyses of light and vision as the products of matter in motion. Using the laws of motion to compare the motions and tendencies of particles of the second element to those of a stone on a sling, Descartes explained how motion travels in rectilinear rays from luminescent heavenly bodies. When these motions strike the surface of a human eye, they cause a sensation of light. This light has a number of characteristic properties that must be accounted for by any explanation, including the facts that it extends from luminous bodies instantaneously in straight lines to any distance; that rays of light can cross paths without interfering with each other; and that they can be diverted by reflection. Considering rays of light to be composed of streams of particles of the second element, Descartes explained all these phenomena in mechanical terms.

La dioptrique (Optics), published in 1637 as one of the Essays printed with the Discourse on Method, contains Descartes' most significant contributions to optics: a mechanical explanation of the phenomena of light; proofs of the laws of reflection and refraction on the basis of mechanical assumptions; and an account of the physics and physiology of vision. He did not claim to know the physical nature of light, but he formulated three different mechanical models—he called them comparaisons (comparisons)—which enabled him to explain the observed properties of light. He used these models to explain the perception of colors, the transmission of light through solid but transparent matter, and the laws of reflection and refraction.

In the first model, he compared light to a stick that enables a blind man to perceive the various objects in his environment by touch alone. This model also makes plausible the claim that the perception of colors results from matter in motion. "You have only to consider that the differences which a blind man notes among trees, rocks, water, and similar things through the medium of his stick do not seem less to him than those among red, yellow, green, and all the other colors seem to us; and that nevertheless these differences are nothing other, in all these bodies, than the diverse ways of moving, or of resisting the movements of, this stick."7

In the second model he appealed directly to his theory of the elements in comparing the rectilinear transmission of light and the passage of light through solid, transparent matter to wine flowing through a vat full of grapes and passing out of a hole in the bottom of the vat:

> Now consider that, since there is no vacuum in Nature, as almost all the Philosophers affirm, and since there are nevertheless many pores in all the bodies that we perceive around us, as experiment can show quite clearly, it is necessary that these pores be filled with some very subtle and very fluid material, extending without interruption from the stars and planets to us. Thus, this subtle material being compared with the wine in that vat, and the less fluid or heavier parts, of the air as well as of other transparent bodies, being compared with the bunches of grapes which are mixed in, you will easily understand the following: Just as the parts of this wine . . . tend to go down in a straight line through the hole [and other holes in the bottom of the vat] . . . at the very instant that it is open . . . without any of

7. René Descartes, Discourse on Method, Optics, Geometry, and Meteorology, trans. Paul J. Olscamp (Indianapolis: Bobbs-Merrill, 1965), p. 67.

those actions being impeded by the others, nor by the resistance of the bunches of grapes in this vat . . . in the same way, all of the parts of the subtle material, which are touched by the side of the sun that faces us, tend in a straight line towards our eyes at the very instant that we open them, without these parts impeding each other, and even without their being impeded by the heavier particles of transparent bodies which are between the two.[8]

In the third comparison Descartes used a tennis ball as a model to derive the known laws of reflection and refraction, thus demonstrating how these fundamental optical phenomena could be incorporated into a mathematized mechanical philosophy. He based his demonstration on the three laws of motion. The law of reflection—that the angle of incidence equals the angle of reflection (both of the angles measured from a line perpendicular to the reflecting surface)—had been known since ancient Greek times. Ptolemy was the first to publish the result. Descartes' contribution was to show how the law followed from his laws of motion when he used the model of the tennis balls for the particles that compose rays of light.

He approached the explanation of the law of refraction similarly. Refraction is the bending of light as it passes from one medium to another. One common example is the apparent bending of an oar when it is partly immersed in water. The law of refraction, in modern notation, states

$$\sin i = n \sin r \,,$$

where $i$ is the angle of incidence and $r$ is the angle of refraction; $n$ is now called the index of refraction. Unlike the law of reflection, the law of refraction had eluded investigators since antiquity. Thomas Harriot (ca. 1560–1621) discovered the law empirically, but his results remained unpublished until the twentieth century. Stimulated by Galileo's use of the telescope, Kepler tried to analyze lenses mathematically, but still failed to find a law. Around 1620, Willebrord Snel (1580–1626) discovered the law empirically. Descartes may have known Snel's work or may have found the result by his own experimental research.

Descartes' proof of the law of refraction followed the same strategy as his proof of the law of reflection. Comparing a particle from a ray of light to a tennis ball, he asked what would happen if the tennis ball struck and penetrated a soft surface like a cloth or water as a model of a ray of light passing from air to water. Once again, he approached the problem by breaking down the particle's

8. Ibid., p. 69.

## Descartes' Analysis of the Reflection of Light

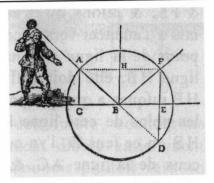

Descartes compared the reflection of light to the reflection of a tennis ball that bounces off the ground. Consider a ball moving from $A$ to $B$ at a constant speed. What happens when it strikes the surface CBE, which Descartes assumed to be perfectly flat and hard? Given the Galilean assumptions about the nature of motion, it is possible to analyze the motion AB into two components, $AH$ and $AC$, moving at right angles to each other. When the ball arrives at $B$, the component $AH$, moving in a straight line at a uniform speed, will be unchanged; therefore, it will travel from $H$ to $F$ in the same amount of time that it traveled from $A$ to $H$. This conclusion follows from Descartes' first law of motion: that each particular part of matter always continues in the same state unless collision with other objects forces it to change its state.

The other component, $AC$, is an example of impact, a phenomenon ruled by the third law of motion. Because in this case the ball is much smaller than the ground it strikes, it bounces off the ground, returning to its original place at the same speed it had before the impact but in the opposite direction. The two new components $HF$ and $EF$ combined will place the ball at F. Now a simple geometrical argument proves that the two triangles $ABH$ and $FBH$ are congruent. Consequently, the angles $ABH$ (the angle of incidence) and $HBF$ (the angle of reflection) are equal.

■ René Descartes, *Discours de la méthode pour bien conduire sa raison et chercher la verité dans les sciences, plus la Dioptrique, les Météores, et la Géometrie qui sont des essays de cette methode* (Leiden: J. Maire, 1637), p. 15.

path into its component motions. As in the case of reflection, Descartes drew a direct analogy between a mechanical model and the phenomenon of refraction in order to show that the optical phenomenon could be derived from the first principles of his natural philosophy.

What epistemological status did Descartes ascribe to these mechanical models? He compared his approach to that of the pre-Copernican astronomers who had constructed their models from various combinations of uniform circular motion without claiming physical reality for them. Descartes assumed that his models must be couched in terms of the fundamental laws of nature and his ultimate terms of explanation, matter and motion. But he stated that he could do better than the astronomers had. He thought he could move be-

## Descartes' Analysis of the Refraction of Light

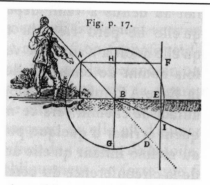

Fig. p. 17.

Once again, Descartes compares the ray of light to a ball, but in the case of refraction, he supposes that the ball penetrates a soft surface. Suppose that the ball traveling from $A$ to $D$ strikes water at point $B$. Suppose also, that when the ball strikes the water it loses half its speed. This ball must pass from $B$ in a straight line, not toward $D$, but toward $I$. The reason for this change of direction is that the water slows the downward component of the motion, so that $EI$ is equal to half of $HB$. But it does not affect the speed of the component $BE$, which is equal to the component $AH$.

The exact quantity of deflection of light as it passes from one medium to another must be determined experimentally. For each medium, these measurements will provide the constant of proportionality— or index of refraction, to use the modern phrase. If that constant is expressed as $n$, then the light is refracted according to the formula $HB/AB = n\ EI/BI$, that is, sin $i =$ $n$ sin $r$, where $i = ABH$ and $r = EBI$.

- René Descartes, *Discours de la méthode pour bien conduire sa raison et chercher la verité dans les sciences, plus la Dioptrique, les Météores, et la Géometrie qui sont des essays de cette methode* (Leiden: J. Maire, 1637), p. 11.

yond hypothetical models and actually establish the truth of his mechanical models by a complex process of experiment and observation.

Having shown—to his own satisfaction—that the properties of light could be incorporated into his mechanical philosophy of nature, Descartes turned to the question of vision, which he approached as a special case of the senses in general. Sensations occur when motions pass from our sense organs through the fibers of the optic nerve to the brain. These motions travel instantaneously, "just as pulling one of the ends of a very taut cord makes the other end move at the same instant."[9] Although these motions produce images in the brain, there is no reason to think that these images resemble the objects that produce the motions.

In order to understand how vision takes place, he began by investigating how images are formed on the back of the eye. Going beyond Kepler's account,

9. Ibid., p. 89.

Descartes based his discussion of the images that form on the back of the eye on observation of dissected eyes either "of a newly deceased man, or, for want of that, of an ox or some other large animal."[10] Following Kepler's strategy of using point-by-point analysis of both the object and the retinal image, Descartes traced light rays from points on the object to points on the surface of the retina at the back of the eye. The rays must pass through several interfaces between different media within the eye, undergoing refraction at each of these surfaces. Using the law of refraction and techniques of tracing rays, Descartes showed how an image of the object is produced on the surface of the retina. He claimed he could then show how this image could be transported to the interior surface of the brain. "And from [the interior surface of the brain] I could again transport it right to a certain small gland [the pineal gland] which is found about the center of these concavities, and which is strictly speaking the seat of the common sense."[11]

Philosophers going back to Aristotle had postulated the existence of the common sense, an organ in which data from all five senses are integrated into a unitary signal that causes conscious awareness. Descartes located the common sense in the pineal gland, which he also thought served as the connection between mind and body. As an aside, he added an explanation of a phenomenon commonly accepted by early modern natural philosophers: "I could even go still further, to show you how sometimes the picture can pass from there through the arteries of a pregnant woman, right to some specific member of the infant which she carries in her womb, and there forms these birthmarks which cause learned men to marvel so."[12]

After deriving the law of refraction, Descartes turned to an analysis of lenses. Although lenses had been used for centuries to correct vision, no one had understood just how they worked. The choice of lenses for spectacles had been simply a matter of trial and error. Using the law of refraction, Descartes was the first to explain how lenses can correct specific defects of vision. An important part of his explanation involved determining the particular curves that focus parallel rays at a single point. Descartes worked out the calculations in his book on geometry, published at the same time as the *Optics*. He applied the same methods to explain the workings of the telescope.

Analysis of the rainbow provided Descartes with a particularly dramatic example by which to demonstrate the power of his natural philosophy. From

10. Ibid., p. 91.
11. Ibid., p. 100.
12. Ibid.

## Descartes' Illustration of the Formation of the Retinal Image

Rays of light coming from points $V$, $X$, and $Y$ travel to the surface of the eye at $B$, $C$, and $D$. They undergo four refractions within the eye before reaching the retina at points $T$, $S$, and $R$. The rays $VR$ and $YT$ cross as they pass through the surface of the eye, and so the image on the retina is inverted.

- René Descartes, *Discours de la méthode pour bien conduire sa raison et chercher la verité dans les sciences, plus la Dioptrique, les Météores, et la Géometrie qui sont des essays de cette methode* (Leiden: J. Maire, 1637), p. 36.

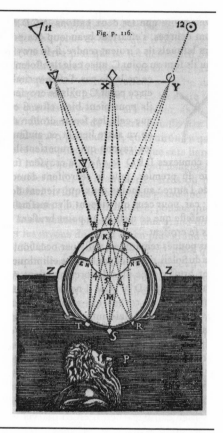

the time of Aristotle, philosophers, astronomers, and natural philosophers had attempted to explain several features of rainbows: their shape, their size, their cause, and the origin of their colors. During antiquity and the Middle Ages, the basic geometry of the rainbow was known, in particular that the rainbow is circular and that the maximum height of the rainbow is 42 degrees as measured by the angle formed by the line from the eye of the observer to the center of the rainbow and the line from the eye of the observer to the top of the rainbow. Rainbows are visible only when the sun is opposite the clouds and rain and only when the sun is fairly low in the sky. Aristotle and a number of medieval thinkers had attempted to explain the size and shape of the rainbow in terms of various combinations of reflections and refractions of sunlight from a mass of clouds and raindrops.

Theodoric of Freiburg (d. ca. 1310) developed innovative ideas about the rainbow based on experimental work. Probably influenced by Alhazen's and

Witelo's point-by-point analysis of light and vision, he thought that the reflections and refractions of sunlight within individual raindrops cause the rainbow. In order to explore the behavior of light within the drops, he used a spherical bowl of water as a model of one raindrop and traced the path of rays of light as they entered the drop, were refracted at its surface, and then were reflected off the back of the drop, undergoing another refraction on exiting the drop. Using this procedure, he determined empirically the angles at which the rainbow is visible and the angles at which particular colors are visible. He could also explain the formation of the secondary rainbow as the result of an additional reflection within drops that are higher in the sky than those that produce the primary rainbow.

Whether or not Descartes was aware of Theodoric's theory, his approach to the rainbow in the *Meteorology* (published at the same time as the *Optics*) is remarkably similar to that of the medieval writer. He began by using a spherical glass vessel to study the behavior of light in individual raindrops. Recounting experiments with a transparent flask, he conducted similar measurements at different places in the sphere of water, obtaining results similar to Theodoric's. Like Theodoric, Descartes traced the path of the rays in the individual drops. He worked out the paths that the light follows in both primary and secondary rainbows, concluding that "the primary rainbow is caused by rays which reach the eye after two refractions and one reflection, and the secondary by other rays which reach it only after two refractions and two reflections."[13] This analysis explains why the secondary rainbow is less bright than the primary one and also why its colors appear in the reverse order. This experimental work enabled Descartes to account for the geometrical properties of the rainbow.

Another question remained: Why does the rainbow exhibit colors? Like virtually all writers since Aristotle—including Descartes' younger contemporaries Francesco Grimaldi (1618–63), Robert Boyle, and Robert Hooke (1635–1703)—Descartes thought that colors result from modifications of white light. Once again using balls as models of the particles of light, he explained color as the result of spin on the particles composing rays of light. When the spin is faster than the forward motion of the particles, the light appears stronger, the strongest color being red; when the spin is slower than the forward motion of the particles, the light appears blue, the weakest color being violet. The refractions that the light undergoes as it enters and leaves the raindrops affect the spin on the particles, thus producing the colors of the rainbow.

When he considered light in the *Principles of Philosophy,* Descartes' con-

13. Ibid., p. 201.

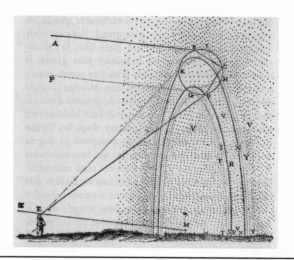

## Descartes' Explanation of the Rainbow

GHK is one drop of water in the rainbow. AB is a ray of light coming from the sun. The light is refracted at B, on the surface of the drop, is reflected at C, and is refracted again at D as it leaves the drop and travels to the observer's eye at E. The path of this light is part of the primary rainbow, on which D lies.

Another ray of sunlight, FG, under-goes two reflections inside the drop at H and I before exiting at K and striking the observer's eye at E. It produces the secondary rainbow on which the point K lies.

■ René Descartes, *Discours de la méthode pour bien conduire sa raison et chercher la verité dans les sciences, plus la Dioptrique, les Météores, et la Géometrie qui sont des essays de cette methode* (Leiden: J. Maire, 1637), p. 251.

cern was largely to incorporate light into his general philosophy of nature and cosmology. Accordingly, he tried to explain how the sun and stars produce light. Dropping the hypothetical approach that characterized his discussion of the three models in the *Optics*, he stated that light is a force or tendency in which the particles comprising these luminous bodies have to recede from the centers about which they revolve. The vortices produce pressure outward from their centers, a centrifugal force. This force, in turn, puts pressure on the particles of the second matter that fill all space. Hence, we see light coming from the centers of the vortices of these heavenly bodies, and we know that, like pressure in a fluid, it is propagated instantaneously.

Subsequent natural philosophers took up questions about light, often starting from the Cartesian account. Descartes' discovery of the law of refraction stimulated a search for proofs based on fundamental principles of motion. Pierre de Fermat (1601–65) sought to deduce the law of refraction from one of the oldest assumptions guiding the study of nature: that nature does nothing in vain. Fermat interpreted this principle to mean that processes would occur in the least time. Somewhat later in the century, Huygens demonstrated how the law followed from his own wave theory of light. In his *Traité de la lumière* (*Treatise on Light*) (1690), he aimed to solve certain difficulties with the Cartesian system. The most notable of these difficulties were why the velocity of light changes as it passes from one medium to another and why rays of light do not interrupt each other when they cross. Huygens envisioned light as a series of waves, emanating from a central disturbance. Although he considered light to consist of waves propagated through a material medium from places on the surface of a luminous body, he did not consider these waves to be periodic (that is, to occur at regular intervals) because the motions of particles on the surface of luminous bodies are not regular. Each point on the spherical wave front becomes the center of a secondary wave. To explain the fact that light travels in straight lines, Huygens assumed that waves are not effective for vision except at their point of tangency with the primary wave, and those tangents are perpendicular to radii—straight lines emanating from the center of the circular wave.

Huygens derived the law of refraction using his wave theory of light. He noted that when a wave front strikes a refracting surface, as when light passes from air to water, the rays that strike the surface first will either speed up or slow down depending on the optical properties of the new medium. This change of motion will affect each ray in turn until the direction of the entire wave front changes. Using simple geometric analysis, Huygens then demonstrated how the sine law of refraction follows from his wave theory of light.

Although Huygens developed a wave theory of light, he denied the periodicity of waves. Isaac Newton thought that rays of light consist of streams of material particles, but he introduced a notion of periodicity in attempting to explain what are now called interference phenomena, such things as the circles of colors formed when water or other fluids are pressed between two pieces of glass. Newton took an innovative approach, deciding to analyze the phenomena quantitatively by measuring the distance between the pieces of glass. He used one flat piece of glass and pressed it against a convex lens. By observing the distance of the rings from their center as he pressed the pieces of glass together, and by clever geometrical analysis based on his knowledge of the radius of curvature of the lens, he was able to calculate the distance between the two pieces of glass at the dark rings and the light rings. He explained the fact

## Huygens' Conception of Light as a Wave Front

In the fluid medium through which light travels, some agitation at *A* creates a wave front that travels successively to *BG*, *dd*, and *CE*. At each point on the wave fronts—*b*, *b*, *b* and *d*, *d*, *d*, etc.— secondary waves develop. These waves are not periodic. The only points on the wave front that produce vision are the ones that are tangential to the main wave front. Hence, light travels in straight lines as, for example, along *ABC* and *AGE*.

■ Christiaan Huygens, *Traité de la lumière, où sont expliquées les causes de*

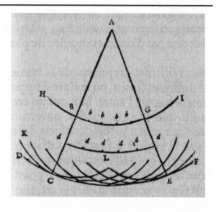

*ce qui luy arrive dans la reflexion et dans la refraction, et particulierement dans l'etrange refraction du cristal d'Islande* (Leiden: Pierre vander AA, 1690), p. 19.

that light is reflected from the thin film at certain diameters and transmitted at others by speculating that the particles of light set up vibrations in the fluid when they strike it. Depending on whether successive particles strike the fluid when the vibration in the fluid is moving toward or away from them, they will be either reflected or transmitted. Newton called these vibrations "fits of easy reflection" and "fits of easy transmission." His theory of "fits" introduced periodicity into his analysis of light, even though, unlike Huygens, he adopted a particulate theory of light.

Until Newton, virtually all natural philosophers agreed with Descartes in supporting the Aristotelian claim that colors result from the modification of white light. Newton overturned this traditional assumption with his famous experiment with light and colors, published in the *Philosophical Transactions of the Royal Society* (one of the earliest "scientific" periodicals) in 1671/72.[14] In this paper he demonstrated that white light consists of irreducible, colored rays.

14. The English did not adopt the Gregorian calendar that had been established by the Roman Catholic Church in the 1582 for about 150 years because they considered the new calendar to be a "Popish plot." Consequently, there were some anomalous aspects to writing dates. In the old Julian calendar, to which the English adhered, the year began on March 25, while according to the Gregorian calendar, the year begins on January 1. Thus, for dates falling between January 1 and March 25, the English used the compound form: the date before the slash is the year according to the Julian calendar, while the date after the slash indicates the year according to the Gregorian calendar.

## Huygens' Demonstration of the Law of Refraction

Let $AC$ be a plane wave front that obliquely strikes the separating surface $AB$ at $A$. As each point on the wave front $KKK$ hits the interface, its forward motion will slow down, so that the wave front $BN$ will bend relative to the original wave front $CB$. The angle of incidence is $EAD$. The angle of refraction is $FAN$. Analysis of the geometry of this figure shows that $\sin EAD = n \sin FAN$, where $n$ is the ratio of the velocities of light in the two media.

- Christiaan Huygens, *Traité de la lumière, où sont expliquées les causes de*

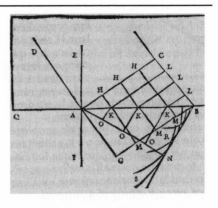

*ce qui luy arrive dans la reflexion et dans la refraction, et particulierement dans l'etrange refraction du cristal d'Islande* (Leiden: Pierre vander AA, 1690), p. 36.

He reported that he passed a beam of sunlight through a prism and projected the resulting spectrum on the far wall of his room, which was about 22 feet (6.7 meters) from the window. He considered the paths of three rays—the two at the edges of the beam and one bisecting the beam as it entered the prism. Although he found that the beam incident on the prism was circular, the spectrum projected on the far wall was about four or five times as long as it was wide, not the circle that one would expect as the image of the sun. The distortion was much greater than the law of refraction predicted. Newton tried rotating the prism and substituting prisms made of different kinds of glass. These changes made no significant difference in the results. He then formed the hypothesis that the white sunlight actually consists of rays of differently colored light, and that passage through the prism refracts each ray differently.

Newton tested this hypothesis by what he called his "crucial experiment." After producing the elongated spectrum in the usual way, he passed a ray of light of one color through a second prism and observed that its color did not change. If the traditional theory that colors result from the modification of white light were correct, the color should change when it passed through the prism. The fact that it did not change indicated to Newton that the original beam of white light consisted of rays of differently colored light, the rays of each color having a characteristic "degree of refrangibility," or what we would call index of refraction, which Newton carefully measured. As a final experi-

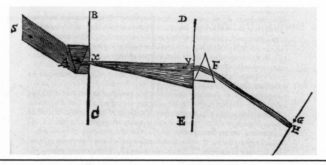

### Newton's Crucial Experiment Demonstrating the Analysis of White Light

Newton closed the shutters of his window and let a beam of sunlight, $S$, enter the room through a small hole. The beam passed through a prism, $A$, which produced a spectrum on a board, $BC$. Newton passed a beam of a single color through a small hole in the board $x$, and the beam travelled to another board, $DE$, where some of it passed through a hole, $y$. This beam passed through a second prism, $F$, which refracted it. The refracted beam traveled to the far wall, landing at $H$. The color of the beam remained unchanged. Newton interpreted this result to mean that white light consists of a mixture of colored beams; passage through the first prism separates it into its component colors. Passing through the prism does not modify the white light because passing the colored beam through the second prism does not change its color.

Newton repeated this experiment with each of the spectral colors, measuring their angles of refraction as they passed through the second prism. He determined that each color has a different angle of refraction. He called this angle the color's "degree of refrangibility." In a further experiment with a converging lens, he recombined all the colored rays and produced a beam of white light.

Newton concluded that passage through the first prism separated the rays of different colors that compose white light because of their respective degrees of refrangibility. Consequently, contrary to long tradition and contemporary claims, Newton stated that colors are not the result of the modification of white light, bur rather that white light is composed of rays of differently colored light.

▪ Isaac Newton, "Mr. Newton's Answer to the Foregoing Letter," *Philosophical Transactions*, no. 85, July 15, 1672, p. 5016.

mental proof of his theory he passed the spectrum of separated colors through a converging lens, which recombined the rays and produced a beam of white light. These experiments resulted in a completely revised theory of colors.

Newton's contemporaries did not immediately understand the significance of his discovery. Instead of attending to the experimental results, they focused on Newton's speculation that the hypothesis that light is composed of particles and the particles of the differently colored rays have different sizes could explain the differences between the differently colored rays. Almost all of his critics—Robert Hooke, Christiaan Huygens, and the Jesuits Ignace-Gaston Pardies (1636–73) and Francis Linus (1595–1675)—argued against his provisional mechanical explanation, without understanding that he had disproven the claim that colors result from the modification of white light. Newton clearly distinguished between an empirical, experimental result and a speculative, mechanical theory, a distinction that few of his contemporaries understood.

Like astronomy, both the science of motion and the science of optics underwent significant changes in the seventeenth century. The Aristotelian concept of motion gave way to a concept based on the principle of inertia, facilitating the application of mathematics to problems that had traditionally remained formulated in basically qualitative terms. The discovery of previously unknown phenomena of light raised new questions about the physical nature of light.

Both disciplines, traditionally considered to be branches of mixed mathematics, applied mathematical methods to physical problems. Linking the science of motion with traditional mechanics, on the one hand, and searching for mechanical models to explain the geometrical properties of light and vision, on the other, involved realigning disciplinary boundaries. A new discipline, mathematical physics, emerged from this process. According to the Aristotelian classification of the sciences, such a discipline would have been considered a contradiction in terms. By the later years of the seventeenth century the existence of this field represented an important change, not only in the content of studies of the natural world, but also in its organization. The Aristotelian scheme fell to the wayside as the new discipline emerged, redefining the methods and content of natural philosophy.

# 6 Exploring the Properties of Matter

*Alchemy and Chemistry*

The study of matter took two different forms in the early modern period. Natural philosophers discussed the properties of matter in general. Alchemists and chemists studied the properties of particular kinds of matter using observational and experimental methods, proposing explanations based both on theories that their predecessors had proposed and on the results of their own experiments and observations. These methods and theories had deep roots, stretching back into ancient times.

Several different activities led to the study of the properties of matter. Medicine, metallurgy, and philosophy each played a role in the development of alchemy and chemistry. Clear boundaries between alchemy and chemistry emerged only in modern times. Indeed, the modern distinction between alchemy (as gold making) and chemistry (as the more general science of the properties, preparation, transformations, interactions, and structure of matter) was not common until the eighteenth century, a fact reflected by the seventeenth-century term "chymistry" and its cognates.

## Seeking Gold and Good Health: Renaissance Transmutations

Paracelsus (1493–1541), whose full name was Philippus Aureolus Theophrastus Bombastus von Hohenheim, inherited these chemical and alchemical traditions along with Hermetic cosmology, but he put these ideas to new uses in the field of medicine. His followers created a chemical philosophy that rivaled the mechanical philosophy in the seventeenth century, influencing the fields of chemistry and medicine as profoundly as Copernicus and Vesalius influenced developments in astronomy and anatomy, respectively. Aggressively critical of traditional medicine, Paracelsus challenged the medical authorities and led a turbulent life as a wandering scholar.

Paracelsus' life followed a repeated pattern of flamboyant success followed by dishonorable failure. His skill as a physician led to public rewards, which he then proceeded to lose when he bit the hand that fed him. For example, in 1527, he went to Basel to treat the famous publisher Johannes Froben, who suffered from an ailment that had resisted all previous treatment. After effecting a cure, Paracelsus accepted an appointment as the municipal physician of Basel, an office which came with the right to lecture at the university. Upon receiving these honors, he challenged the faculty of the university by proclaiming that he would lecture for two hours every day on his own philosophy of nature and theory of disease, which he claimed to have based on his own observations, rather than on the texts of Hippocrates and Galen that formed the basis of the medieval medical curriculum. Moreover, rather than lecturing in Latin, the official language of academia, he lectured in his native Swiss-German. As a climax to this challenge, he threw the volumes of Avicenna's authoritative *Canon of Medicine* into the students' bonfire on St. John's Eve. Within a short time, the university revoked his right to lecture, and he faced charges of malpractice. After the death of Froben and continual controversies in which he insulted various officials, he left town in a hurry, leaving behind his possessions and the manuscripts of his books. Episodes like those at Basel were repeated throughout his life.

Despite a lifetime of turmoil and controversy, Paracelsus formulated ideas that revolutionized medicine. He believed that chemistry provides the key to understanding both nature and medicine. He argued that Aristotle, Galen, and Avicenna—the three authorities of the universities' medical curriculum—were each ignorant of chemistry, grounds for dismissing their ideas entirely. In place of traditional medicine, he proposed a chemical philosophy that incorporated Hermetic ideas about the correspondence between the macrocosm and microcosm, a relationship which underpinned the unity that God had implanted in the cosmos. For Paracelsus, the microcosm is the human being:

> All this you should know exists in man and realize that the firmament is within man, the firmament with its great movements of bodily planets and stars which result in exhalations, conjunctions, oppositions and the like, as you call these phenomena as you understand them. Everything which astronomical theory has searched deeply and gravely by aspects, astronomical tables and so forth,—this self-same knowledge should be a lesson and teaching to you concerning the bodily firmament. . . . You are aware that the earth exists solely for the purpose of bearing fruit and for the sake of man. With the same logic, the body also exists solely for the same reason. Thus from within the body grows all the food which is to be used by the members that belong to the body. These grow like the fruit of the earth.[1]

1. Paracelsus, *Volumen medicinae paramirum*, quoted in Allen G. Debus, *The Chemical Philosophy: Paracelsian Science and Medicine in the Sixteenth and Seventeenth Centuries*, 2 vols. (New York: Science History Publications, 1977), 1:53.

Emanations from the heavenly bodies cause the correspondence between the macrocosm and the microcosm. The physician needs to understand this relationship as well as the secrets hidden in earthly things. Instead of the books of the ancients, he should rely on his own experience and that of others: "A Physitian ought not to rest only on that bare knowledge which their Schools teach, but to learn of old Women, Egyptians, and such-like persons; for they have greater experience in such things than all the Academicians."[2] This experiential knowledge included prayer, faith, and imagination, as well as straightforward empirical observation.

Paracelsus thought that the main processes by which the world and the human body work are alchemical. Accordingly, he gave an alchemical account of the Creation, claiming that the principle of all generation is separation, which, along with distillation, was one of the key alchemical processes. He adopted a modified version of the Aristotelian theory of the four elements, supplementing them with three "principles"—salt, sulfur, and mercury—which endow ordinary matter with its characteristic properties. The three principles are not the same as the ordinary substances having those names; rather, they are the bearers of qualities into ordinary substances. Salt introduces solidity; sulfur introduces odor, color, and flammability; and mercury, volatility. A simple experiment seemed to bear out the presence of all three principles in a green twig. When the green twig burns, it produces ash (salt), flame (sulfur), and smoke (mercury), each of which the process of burning releases from the twig.

Unlike Hippocratic and Galenic physicians, who regarded disease as systemic, resulting from an imbalance of the four humors, Paracelsus thought that diseases arise from causes that are outside the person and that affect specific parts of the body. Because the various parts of the body correspond to various heavenly bodies, the physician can find cures by understanding the correspondences between the stars and the parts of the body and between the stars and the metals, minerals, and plants on the earth. Astrology thus permeated Paracelsian medicine. Alchemy also played a central role. According to Paracelsus an *archeus*, a sort of inner alchemist that can apply the medications prescribed to cure specific areas of disease, regulates each organ of the body. On the basis of this theory of correspondences, Paracelsus introduced the use of chemical remedies, rejecting the traditional reliance on herbal concoctions. He initiated the use of mercury compounds to treat syphilis, a devastating disease that appeared in Europe after Columbus returned from the New World.

2. Paracelsus, *Of the Supreme Mysteries of Nature*, quoted in Debus, *Chemical Philosophy*, 1:54.

A saying popular at the time warned, "One night with Venus yields three years with Mercury."

After Paracelsus' death, a number of his followers proposed a new, chemical philosophy to replace the works of Aristotle and Galen, which they regarded as theologically unsound and philosophically useless. Advocates of this chemical philosophy compared Paracelsus' contribution to medicine to that of Copernicus in astronomy and of Luther and Calvin in theology. Like these other men who had looked back in order to move forward, the chemical philosophers believed that they could restore Adam's prelapsarian, pristine knowledge of nature by finding wisdom in both the Old Testament and the Hermetic writings. They frequently invoked the metaphor of God's two books: the book of God's word (the Bible) and the book of God's work (the created world).

For the Paracelsians, chemistry unlocks the secrets of natural philosophy. Not only did they seek an educational reform based on the study of the Bible and the study of nature, but they even interpreted the account of Creation in Genesis (the first book of the Bible) and the end of the world as prophesied in Revelation (the last book of the Bible) in chemical terms. They believed that a new chemical philosophy would provide insight into God's works and would also serve as the foundation for a new, chemical approach to medicine based on the correspondence between the macrocosm and the microcosm. Like the mechanical philosophers who were seeking to replace Aristotelianism, they presented their philosophy as the foundation for a new natural philosophy.

The English Paracelsian and chemical philosopher Robert Fludd (1574–1637) provides a vivid example of the chemical philosophy in action. He called for reforms in medicine and natural philosophy, arguing that knowledge of the occult sciences rather than the tired philosophy of Aristotle should serve as foundations for both of these subjects. He considered mathematics to be the key to these sciences, but not the simple mathematics of geometry, astronomy, and optics. Instead, he sought deeper meaning from mathematics, a meaning based on a Neoplatonic and Pythagorean approach that leads to a more profound understanding of the symbolic meanings embedded in the created world.

The controversy surrounding Fludd's endorsement of the Paracelsian "weapon salve" illustrates both the cosmological commitments of this new philosophy and its differences from both Aristotelianism and the mechanical philosophy. According to a theory attributed to Paracelsus, the weapon salve is an ointment that can cure a wound by being applied to the weapon that inflicted the wound rather than to the wound itself. Composed of a witches' brew of disgusting ingredients, including blood from the wound (still warm), fat, and moss taken from the skull of a hanged criminal, it can act over vast distances. According to Fludd and other Paracelsians, the efficacy of the weapon salve is a product of the relationship between the macrocosm and the microcosm, a relationship which produces an occult sympathy between the wounded person and the weapon. Since both the weapon and the wound contain traces of the wounded person's blood, both correspond to the same heavenly bodies; and this nexus of correspondences accounts for the possibility of action-at-a-distance.[3]

The sympathies between cosmically related objects and substances could account for a host of strange phenomena, such as the story of a "certaine Lord or Nobleman of Italy" who having

> by chance lost his nose in a fight or combate, this party was counselled by his Physicians to take one of his slaves, and make a wound in his arme, and immediately to ioyne his wounded nose to the wounded arme of the slave, and to binde it fast, for a season, untill the flesh of the one was united and assimilated to the other. The Noble Gentleman got one of his slaves to consent, for a large promise of liberty and reward; the double flesh was made all one, and a collop or gobbet of flesh was cut out of the slaves arme, and fashioned like a nose unto the Lord, and so handled by the Chirugion that it served for a naturall nose. The slave being healed and rewarded, was manumitted, or set at liberty, and away he went to Naples. It happened, that the slave fell sicke and dyed, at which instant, the Lord's nose did gangrenate and rot; whereupon the part of the nose which hee had of the dead man, was by the Doctors advice cut away, and hee being animated by the foresaid experience, followed the advice of the same Physician, which was to wound in like manner his owne arme, and to endure with patience, till all was complete as before. He with animosity and patience, did undergoe the brunt, and so his nose continued with him until his death.[4]

A controversy about the weapon salve raged around Fludd in the 1630s when he rose to its defense against accusations that it was the deceitful art of the devil. An English clergyman and Aristotelian, William Foster, wrote a tract attacking Fludd's endorsement of the salve, saying that it must be the work of the devil and that Fludd must therefore be a witch. Foster noted that the

---

3. For a fictional account of the weapon salve and other unusual aspects of early modern natural philosophy, see Umberto Eco, *Island of the Day Before*, trans. William Weaver (New York: Penguin, 1995).

4. Robert Fludd, *Doctor Fludds Answer unto M. Foster; or, The Squeezing of Parson Fosters Sponge, Ordained by Him for Wiping Away of the Weapon-Salve* (1631), quoted in Debus, *Chemical Philosophy*, 1:247.

weapon salve is not mentioned in the Bible, so it cannot be of divine institution. Moreover, it violates a basic principle of Aristotelian physics that there can be no action-at-a-distance, so it cannot be a product of nature. The only alternative, claimed Foster, is that it must be the work of the devil. Foster accused Paracelsus of dabbling in magic and witchcraft. "The Author of this Salve, was Philippus Aureolus Bombastus Theophrastus Paracelsus. Fear not Reader, I am not conjuring, they are onely the names of a Conjurer, the first inventer of this Magicall ointment."[5] Foster entitled his tract *Hoplocrisma-spongus; or, A Sponge to Wipe Away the Weapon-Salve*, and nailed two of the title pages to the door of Fludd's house in the middle of the night. Undaunted, Fludd wrote a reply to Foster's criticisms in *Doctor Fludds Answer unto M. Foster; or, The Squeezing of Parson Fosters Sponge, Ordained by Him for Wiping Away of the Weapon-Salve*, in which he rebutted Foster's claim that the weapon salve was diabolical in origin. Fludd laid out the basic principles of his chemical philosophy of nature to explain the weapon salve.

A number of other natural philosophers addressed the question of the weapon salve. For example, Daniel Sennert (1572–1637), whose matter theory fused Aristotelian corpuscularianism with alchemy, argued that to the extent that the weapon salve worked, its success resulted from keeping the wound clean. After describing the weapon salve and dismissing the explanations that involved some kind of action-at-a-distance or the activities of the *anima mundi* (world soul), Gassendi gave an extended atomistic explanation of the salve. He claimed that the blood-soaked sword leaves a trail of atoms, which the curative corpuscles of the salve follow back to the wound.

> Just as, when someone conveys a burning firebrand through the air, there remains a long trail of smoke; so when the sword . . . is transferred, there can remain a trace of an insensible vapor through the air, incited by the blood of the wound, which by its continuation intervenes between the wound and the salve. And lest you think this entirely ridiculous, consider through how much space the odors of things are diffused, especially those with a powerful smell . . . Indeed, think how, when an indiscreet hare or deer flees running rapidly from dogs and is chased in bending curves, a most subtle vapor, imperceptible to our senses, but nevertheless sensible to the dogs, is diffused and remains in the air. For if nothing remained, how could the dogs find it?[6]

5. William Foster, *Hoplocrisma-spongus; or, A Sponge to Wipe Away the Weapon-Salve* (1631), quoted in Debus, *Chemical Philosophy*, 1:281.

6. Pierre Gassendi, *Syntagma Philosophicum*, in *Opera Omnia*, 6 vols. (1658; facsimile repr., Stuttgart–Bad Cannstatt: Friedrich Frommann Verlag, 1964), 1:457.

Gassendi never questioned the efficacy of the salve but sought instead to demonstrate that it, like everything else in the universe, could be explained in atomistic terms.

Despite the excesses of some of the ideas that the chemical philosophers proposed, their theories, particularly their theories of matter, exerted a strong influence on many later developments in chemistry. Their proposals for the reform of the universities, involving the replacement of the Aristotelian- and Galenic-based curriculum by one based on the chemical philosophy, as well as their utopian schemes, indirectly influenced the formation of the Royal Society of London later in the seventeenth century.

## Explaining Matter: Early Modern Chymistry

Quite apart from and in addition to the debate about the weapon salve, Paracelsianism deeply affected thinking about chemistry in the century that followed. One of the most influential chemical philosophers, the Belgian Catholic Joan Baptista Van Helmont (1579–1644), took Paracelsian chemistry to the next level. Although deeply influenced by Paracelsus, he did not follow him blindly and freely criticized many of Paracelsus' ideas. He did, however, emphasize the importance of chymistry (remember that the distinction between alchemy and chemistry did not exist at this time) and its close link with medicine. Van Helmont developed his own philosophy of nature. His work surpassed that of his predecessors in the development of both theory and experimental methods.

Rejecting Aristotle and Galen as heathens, Van Helmont insisted that God created the world and is the ultimate source of motion. Arguing from the order of Creation in Genesis, he rejected the idea that the heavenly bodies can cause motions on earth or have a direct influence on humans. He thought that each person and each organ of the human body has a divine source of motion, which he called "Blas." Accordingly, he argued against any causal relationship resulting from the correspondence between the macrocosm and the microcosm, a relationship that had played a central role in the thinking of Paracelsus and the later chemical philosophers. Van Helmont's formulation of a philosophy of nature thus started from theological considerations.

For Van Helmont, chymistry provided the key to both medicine and natural philosophy. He developed a theory of matter, basing many of his conclusions on experimental evidence. Rejecting the Aristotelian theory of the four elements, he turned to Scripture as a guide. He noted that on the second day of Creation, "God said, Let there be a firmament in the midst of the waters, and let it divide the waters from the waters. . . . And it was so."[7] On the basis of this passage, Van Helmont maintained that water is the single element. The book of Genesis mentions neither fire nor air, and he thought that he had discovered quantitative, experimental evidence to prove that all terrestrial substances ultimately consist of water.

> I took an Earthen Vessel, in which I put 200 pounds of Earth that had been dried in a Furnace, which I moystened with Rainwater, and I implanted therein the Trunk or Stem of a Willow Tree, weighing five pounds; and at length, five years being finished, the Tree sprung from thence, did weigh 169 pounds and about three ounces. But I moystened the Earthen Vessel with Rain-water, or distilled water (always when there was need) and it was large, and implanted into the Earth, and least [*sic*] the dust that flew around should be co-mingled with the Earth, I covered the lip or mouth of the Vessel, with an Iron-Plate covered with Tin, and easily passible with many holes. I computed the weight of the leaves that fell off in the four Autumnes . At length I again dried the Earth of the Vessel, and there were found the same 200 pounds, wanting about two ounces. Therefore 164 pounds of Wood, Barks, and Roots, arose out of water onely.[8]

Van Helmont interpreted the experiment as demonstrating that all the different substances that distillation could extract from the tree—oils, wood, alcohol, salts—originated from water. The quantitative approach evident in this experiment characterized much of Van Helmont's chymical research. His assumption that "no substance is to be annihilated by the force of nature, or art [human technique]," and similarly none is created, underpinned his use of quantitative methods.[9] Consequently, he believed that the weight of substances remains constant in chemical changes.

That all solid, earthly, and even animal and vegetable matter can be reduced to salts, which can then be dissolved into waters of one kind or another, reinforced the results of the tree experiment. Different kinds of matter can be produced from water because of the presence of seminal principles or seeds, which God created at the beginning of the world. Ferments—directing forces coming from God—control the development of the seeds, forming the basic

---

7. Gen. 1:6, 7 (King James version).

8. Jan Baptiste Van Helmont, *Oriatrike, or Physick Refined. The Common Errors therein Refuted, and the Whole Art Reformed & Rectified. Being a New Rise and Progress of Phylosophy and Medicine, for the Destruction of Diseases and Prolongation of Life*, quoted in Debus, *Chemical Philosophy*, 2:319.

9. Ibid., p. 329.

structure of Van Helmont's vitalistic view of nature. Van Helmont also rejected the Paracelsian "principles"—salt, sulfur, and mercury—as constituents
of mixed substances. Although analysis by fire or distillation often produced
these three kinds of substances from bodies, Van Helmont argued that they
were not present before the application of heat, but are in fact artifacts of the
process itself.

Alchemy played a role in Van Helmont's thinking. He claimed to have
succeeded in transmuting mercury into pure gold by using the philosophers'
stone. He also claimed that there exists a universal solvent, the *alkahest*, a
liquid "which resolves every visible Body into its first matter, the power of the
Seeds being reserved."[10]

## Mechanizing Chymistry

Van Helmont's theories and methods influenced many subsequent chymists
and natural philosophers, most notably Robert Boyle. Boyle learned about
Helmontian chymistry from George Starkey (1628–65). Born in Bermuda and
educated at Harvard, Starkey was a practicing alchemist. He moved to London in 1650, where he set up an alchemical laboratory and wrote alchemical
treatises under the pseudonym Eirenaeus Philalethes. Both Boyle and Newton
knew these treatises and incorporated many of the ideas contained in them
into their own thinking. Although Boyle came to know Starkey, who introduced him to chymistry and laboratory practice, he never realized that Starkey
had written the pseudonymous treatises. In the 1640s, the young Boyle, son
of a wealthy Anglo-Irish family, had been concerned primarily with moral and
spiritual issues, having, at best, an armchair interest in natural philosophy. A
sudden shift of attention to natural philosophy, and particularly chymistry,
followed his encounter with Starkey towards the end of 1650. From that point
on, Boyle's writings contain discussions of ideas and specific procedures clearly
borrowed from conversations with Starkey and from Starkey's laboratory note-
books. Starkey's Helmontian outlook deeply influenced Boyle and remained
with Boyle for the rest of his life.

In addition to introducing Boyle to chymistry, Starkey stimulated Boyle's
growing interest in alchemy. Boyle's voluminous publications and unpublished
manuscripts are full of references to alchemy, alchemists, and the philosophers'
stone. Boyle claimed to have witnessed several transmutations, and he contributed money to support alchemical activities in England, Europe, and as

10. Ibid., p. 326.

far away as Turkey. He believed in the efficacy of alchemical procedures and devoted much time and money to his quest for the philosophers' stone. Many seventeenth-century natural philosophers shared Boyle's devotion to alchemy. When Boyle died, Isaac Newton wrote to John Locke (1632–1704), Boyle's literary executor, for a recipe for a mysterious red earth that would supposedly transmute base metals into gold. Newton claimed that Boyle had told him about a recipe for this red earth some years before, but when Newton followed the recipe, he had not succeeded in producing a transmutation. Newton suggested that perhaps Boyle, in the interests of alchemical secrecy, had omitted a crucial step. Now that Locke had access to all of Boyle's papers, Newton hoped that Locke could send him the complete recipe. That the two heroes of the Enlightenment of the eighteenth century engaged in a serious correspondence about an alchemical recipe for gold making left by the so-called father of modern chemistry should cause us to stop in our tracks and reexamine a number of received assumptions!

One of Boyle's most famous treatises, *The Sceptical Chymist* (1661), in which he attacked the traditional elements and principles, bears the marks both of Starkey's laboratory practices and of Van Helmont's approach to chymistry. Boyle argued that experiments and observations do not provide solid evidence for either the three Paracelsian principles or the four Aristotelian elements. Different procedures can analyze a given substance into various different components, and not always the same number of them, as the theories of elements and principles presume. Some substances, such as gold, are extremely stable and do not succumb to chymical analysis at all. Although a burning twig produces smoke, flame, and ash—corresponding to the Paracelsian mercury, sulfur, and salt—distilling the same twig rather than burning it in a open fire produces an entirely different set of components, including oil, spirit (alcohol), and water.

Boyle's definition of "element" has led many commentators and the authors of many textbooks of chemistry to proclaim him as "the father of modern chemistry." But their attribution of fatherhood to Boyle results from an incomplete reading of this passage:

> I now mean by Element, as those Chymists that speak plainest do by their Principles, certain Primitive and Simple, or perfectly unmingled bodies, which not being made of any other bodies, or of one another, are the Ingredients of which all those call'd perfectly mixt Bodies are immediately compounded, and into which they are ultimately resolved: *now whether there be any one such body to be constantly*

*met with in all, and each, of those that are said to be Elemented bodies, is the thing I*
*now question.*[11]

This passage gives the lie to Boyle's paternity. Reading the entire passage quoted
here makes clear that, far from basing his chymistry on a modern concept of
element, Boyle questioned whether any such substances exist. He reinforced
this conclusion in the very next paragraph:

> By this state of the controversie, you will, I suppose, Guess, that I need not be so
> absurd as to deny that there are such bodies as Earth, and Water, and Quicksilver,
> and Sulphur: But I look upon Earth and Water, as component parts of the Uni-
> verse, or rather of the Terrestrial Globe, not of all mixt bodies. And though I will
> not peremptorily deny that there may sometimes either a running Mercury, or a
> Combustible Substance be obtaine'd from a Mineral, or even a Metal; yet I need
> not Concede either of them to be an Element in the sence above declar'd.[12]

Among the many experiments Boyle performed, he repeated Van Helmont's
experiment with the willow tree and with other plants to prove that the Aristotelian four elements are not truly elementary, but can be converted one into
another.

If Boyle's introduction to chymical concepts and procedures came largely
from Starkey, his general approach to natural philosophy had other origins. He
firmly embraced a version of the mechanical philosophy (Boyle actually coined
the term), which he called corpuscularianism. Although Boyle's corpuscular
philosophy permeated most of his writings, he stated it most systematically
in *The Origin of Forms and Qualities* (1666). His philosophy of nature bore
the imprint of both Gassendi and Descartes, although he remained agnostic
on key issues such as whether matter is infinitely divisible and whether a vacuum can exist in nature. He shared the mechanical philosophers' fundamental
assumption that matter and motion can explain all physical phenomena. He
claimed that all bodies are composed of "one Catholick or Universal matter"
that is "extended, divisible, and impenetrable."[13] In order to account for the
diversity of bodies and for their qualitative changes, he added a second fundamental principle, motion.

11. Robert Boyle, *The Sceptical Chymist; or, Chymico-Physical Doubts & Paradoxes, Touching
the Spagyrist's Principles Commonly call'd Hypostatical, As They Are Wont to Be Propos'd and De-
fended by the Generality of Alchymists* (1661), in *The Works of Robert Boyle,* ed. Michael Hunter and
Edward B. Davis, 14 vols. (London: Pickering and Chatto, 1999), 2:345. My italics.

12. Ibid.

13. Robert Boyle, *The Origine of Formes and Qualities (According to the Corpuscular Philosophy), Illustrated by Considerations and Experiments,* in Boyle, *Works,* 5:305.

Matter is divided into small particles that differ from each other only in size and shape. Boyle held a hierarchical theory of matter. He called the smallest particles *minima naturalia*. Although God, in his omnipotence, could divide them further if he so chose, in nature they are scarcely ever divided. These *minima* form microscopic clusters, in which they are bound very closely together. Some clusters remain stable, although they can, in principle, be divided. Substances like gold that resist decomposition are composed of such stable clusters. The configurations of the clusters produce the properties of macroscopic substances. These first-level clusters come together to form larger clusters that are less stable. One interesting feature of *The Origin of Forms and Qualities* is Boyle's attempt to interpret all of the key Aristotelian concepts in corpuscular terms. Accordingly, he considered the Aristotelian concepts of substance, form, qualities, generation, and corruption and reinterpreted each of them in terms of matter and motion. He reinforced Gassendi's and Descartes' program by applying the mechanical philosophy to the content of a specific enterprise, chymistry.

Boyle claimed that the configurations of complex particles and the rearrangement of their parts produce the properties of chymical substances and reactions. He devoted many treatises to demonstrating that chymistry provides some of the best illustrations for the mechanical philosophy. Indeed in the second and much longer part of *The Origin of Forms and Qualities*, he described the analysis and synthesis of various substances and translated all the chemical processes he observed into corpuscular terms.

Boyle's most noted project concerned the nature of air and vacuum. He performed an extensive series of experiments with the newly fabricated air pump to prove that the properties of air—most notably its "spring"—could be explained in mechanical terms. Observing the compression and expansion of air, he hypothesized that air consists of particles that are like little springs—like curly strands of wool. He placed a mercury barometer inside the receiver of a pump and demonstrated that the height of the mercury in the barometer varies in proportion to the pressure of the air in the pump. As he pumped air out of the receiver, thus lowering the pressure, the column of mercury fell. When he let the air back in, the column of mercury rose. These observations led him to conclude that the weight of the air, not the *horror vacui*, causes the column of mercury to be suspended in the tube of the barometer.

Although Boyle was an excellent laboratory chymist, his project to incorporate chymistry into the mechanical philosophy did not prove fruitful for chemistry. Despite knowing many facts about the behavior of chemical substances, neither Boyle nor anyone else at the time knew general principles or laws at the chemical level of analysis. No one really knew how to reduce chemical phenomena to physics. The project was clear to some, but the means to accomplish it did not yet exist. Others denied its possibility. Equally, Boyle's claims to the contrary notwithstanding, it was difficult to see how the mechanical philosophy could actually help the laboratory chymist in his explorations of the properties of chemical substances.

The traditional Paracelsian principles proved to be more useful for furthering chemical knowledge, particularly for the understanding of combustion. A Paracelsian who emphasized the practical side of chemistry and sometimes served as advisor to several German courts, Johann Joachim Becher (1635–82), believed, like Paracelsus, that chemistry is the key to medicine. He thought that metals are composed of three different kinds of earth, one of which is the principle of combustibility. Borrowing a term from Van Helmont, he called this substance phlogiston. Georg Ernst Stahl (1660–1734), a German physician who held positions in both universities and courts, more fully articulated the theory of phlogiston. He opposed mechanical philosophers like Boyle and followers of Newton who wanted to reduce chemical phenomena to physics. Instead, he emphasized the study of the chemical properties of substances.

Stahl thought that water and earth are the two fundamental substances, but he divided earth into three principles: metallic or mercurial earth, which gives metals brightness and malleability; vitrifiable or glassy earth, which makes substances capable of being melted and forming glassy materials; and sulfureous earth or phlogiston, which endows substances with the capacity to burn. These earths could not be isolated, and they endowed substances with their particular chemical properties. Stahl explained both combustion and calcinations (phenomena like rusting) as processes in which phlogiston is lost. The process of calcination is reversible. Heating a calx (the product of calcination) in the presence of phlogiston restores a metal to its original, metallic state. The theory of phlogiston could explain why a candle ceases burning in a closed container: the air in the container becomes saturated with phlogiston and can absorb no more. The reason the atmosphere does not become saturated with phlogiston is that plants absorb phlogiston. Thus, Van Helmont's willow tree grew because it absorbed phlogiston from the air as well as water through its roots.

Further discussions of the phlogiston theory coupled with precise quantitative methods later in the eighteenth century led both Joseph Priestley (1733–1804) and Antoine-Laurent Lavoisier (1743–94) to the discovery of oxygen and the formulation of the modern concept of the chemical element. But that is a story for another day.

Early modern chymistry does not fit neatly into the categories of modern scientific disciplines. With roots in diverse aspects of alchemy, medicine, metallurgy, the Hermetic tradition, and the mechanical philosophy, the study of the properties of matter did not at this time develop into a unified science based on a shared set of assumptions. Modern categories, modern disciplines, and new assumptions are the products of a later age.

# 7 Studying Life

## Plants, Animals, and Humans

Animals, plants, and human beings occupy a central place in the natural world. The study of living things in the early modern period did not fall within one science or a single discipline. Indeed, the term "biology" did not exist before the nineteenth century. Instead, scholars, physicians, and natural philosophers approached the study of living things from several different perspectives. The explorations of regions beyond Europe led to the discovery of previously unknown flora and fauna, challenging the veracity of traditional literary sources and leading to a search for accurate descriptions of the newly discovered creatures. At the same time, naturalists intensively studied local European plants and animals. Medical writers focused on the pharmaceutical virtues of hitherto unknown plants as well as the identification of the plants described in the classical sources. Developments in anatomy led to discussions of the function of bodily parts, thereby promoting a reexamination of traditional physiology. And natural philosophers sought to define human nature by discussing the seemingly eternal question of the distinction between humans and animals, a problem exacerbated by the rise of the mechanical philosophy and the perceived threat of materialism. Theological questions informed each of these areas of study.

### Echoing Aristotle: The Kinds of Plants and Animals

Natural history stretches back to the works of Aristotle, Pliny, and Dioscorides. Like many others areas of study, natural history experienced a revival as a result of the work of the Renaissance humanists. A somewhat unexpected result of the humanists' recovery of ancient texts—a fundamentally literary activity—was a new emphasis on observation as the authoritative source for knowledge of the natural world. Editing the text of Pliny's *Natural History*, the humanist Niccolò Leoniceno (1428–1524) discovered numerous scribal errors and mistranslations of earlier Greek writings in the Roman naturalist's text. His criticism of a classical authority provoked a heated debate with other scholars. Despite attempts by Ermolao Barbaro (1454–93) to correct the mistakes in Pliny's text, other humanists concluded that only by observing the plants he described and comparing his descriptions with both the specimens and the descriptions of other writers could they determine the accuracy of Pliny's descriptions. In this way, a project that started as a purely literary study led to the beginnings of a natural history based on direct observations of plants and animals. Although Leoniceno advocated direct observation enthusiastically, his focus was more on correcting the errors he found in Pliny and Arabic writings on medicine than in writing a new account of plants. In his day ancient texts formed the basis of the medical curriculum; hence, he aimed to remove errors in those sources rather than to create a new, empirical natural history.

Physicians and medical writers also played an important role in reviving the study of natural history. Because physicians needed to be able to judge the competency of apothecaries, they required knowledge of the ingredients of drugs, most of which came from plants. Accordingly, starting in the sixteenth century, Italian medical faculties introduced the study of Dioscorides' *Materia medica*, which Barbaro had translated from Greek and which was published posthumously in 1499. Although Dioscorides' book had been known in Latin in the Middle Ages, before the advent of printing the reproduction of the illustrations with any degree of accuracy remained impossible. In printed editions, illustrations of the plants that Dioscorides described became an important feature of the book after the middle of the sixteenth century, greatly enhancing its utility. To enable physicians to recognize these plants, the medical faculties in Padua and Pisa established medical gardens, where students could observe medically important plants directly.

Conrad Gessner (1516–65) based his detailed descriptions of plants and animals on careful observation as well as folklore, mythology, and symbolic representation. Intending readers to use his *Historia animalium (History of Animals)* (1551) as a dictionary for reference rather than a discursive account of natural history, Gessner listed the animals in alphabetical order. For each animal he supplied several sorts of information: the names of the animal in ancient and modern languages; its geographical distribution, regional differences, and morphology (structure); its behavior; its character, ingenuity, vices and virtues, and sympathies and antipathies; its use as food; its medical uses; and its philological aspects. The last category referred to its name in various languages, metaphors associated with it, and most importantly its symbolic meaning. Usually Gessner determined the animal's symbolic meaning by quoting from the 1531 *Emblemarum liber (Book of Emblems)* by Andrea Alciati (1492–1559). In this book, Alciati printed symbolic pictures on various topics. A poem or epigram accompanied each emblem. The emblems depicting various animals often referred to the traditional folkloric or religious meaning associated with them. In the century following the publication of Gessner's book, emblematic natural history became popular.

Ulisse Aldrovandi (1522–1605), a professor of medicine, taught medical natural history at the University of Bologna, where he established a botanical garden. Aldrovandi maintained a wide-ranging network of correspondence with other scholars and collectors throughout Europe, including Gessner. He established a large collection of natural specimens and other curiosities, which became one of the first natural history museums. He wrote many encyclopedic volumes on natural history, many of which appeared in print only after his death. Where Gessner had listed the animals in his book alphabetically, Aldrovandi categorized them by the shapes of their feet.

In addition to writing encyclopedic works about living plants and animals, Renaissance naturalists wrote works about objects dug out of the ground. The word "fossil" did not refer to organic remains at this time, but simply meant things "dug out of the ground." Accordingly, Gessner's treatise *De omni rerum fossilium genere, gemmis, lapidibus, metallis, et huiusmodi* (*A Book on All Kinds of Fossil Objects, Gems, Stones, Metals and Things of This Kind*) (1565) described a wide variety of objects including gems, crystals of minerals, old coins, buttons, and objects resembling living things. Like his works on plants and animals, Gessner's book on fossils had a strong literary and philological character, as well as being purely descriptive. Both Aldrovandi and Gessner made extensive use of woodcut illustrations to identify the objects they described. Thinking in terms of astrological and Neoplatonic cosmology, Renaissance naturalists tended to explain the origin of these "formed stones" by appealing to the influence of the heavens, the correspondences governing the world, and the Aristotelian theory that stones and metals grow in the earth.

By the sixteenth century, naturalists loosened their bonds to ancient texts and became more interested in establishing a catalog of nature, one that included plants and animals unknown to the ancients. They developed specialized tools for expediting and sharing their observations. Prominent among these tools were herbaria—devices for preserving dried plants between sheets of paper—and botanical gardens that went beyond pharmaceutically useful plants. Herbaria enabled collectors to share specimens and to observe them outside the field setting. Gaspard Bauhin (1541–1613), the first professor of anatomy and botany at the University of Basel, wrote a number of important herbals, emphasizing the careful observation of plants quite apart from their medical uses. Naturalists who had examined new plants communicated their findings with one another either by traveling to share materials in person or by sending each other dried samples of plants. In this way, a community of naturalists arose among collectors who shared their observations and developed precise ways of describing and illustrating the plants that they observed. These herbalists added many descriptions of local plants to their natural histories. The invention of printing made the use of illustrations in books on natural history possible, just as it had in anatomy. In addition to describing a plant's structure, Renaissance naturalists gave detailed accounts of its development and life cycle, the places where it grew, and its various uses, especially medical. Sometimes their descriptions even included recipes for cooking their specimens.

Desire for spices and for medically useful plants motivated many of the expeditions to regions outside of Europe—to Asia and to the New World. Explorers and medical writers often brought back samples of hitherto unknown plants and animals from these distant places. The exotic character of these specimens challenged the classical authorities. Early natural histories of animals had drawn heavily from classical sources, but they came to be modified in the light of creatures observed in distant places. One of the early accounts of the animals of the New World, the *Natural and Moral History of the Indies* by José de Acosta (1540–1600), followed Aristotle's basic format. Nevertheless, Acosta was willing to criticize and correct Aristotle when he observed creatures that did not fit into the Aristotelian scheme. Insisting on basing his descriptions on his own observations, Acosta affirmed that the intention of his work was to glorify the Author of all nature, a theme that would become increasingly prominent in the natural history of the seventeenth century.

Philip II of Spain sent the physician Francisco Hernández (1517–87) to New Spain (Mexico) to report on the medicinal qualities of the plants there. Hernández wrote about both plants and animals, basing much of his account on the knowledge of indigenous people as well as his own observations. His nomenclature incorporated local Nahuatl names as well as Latin and Spanish names. He often compared the plants and animals that he saw in the New World to those known in Spain. Although much of his work and many of the illustrations were lost or remained unpublished during his lifetime, the information that he managed to transmit to Europe was eagerly assimilated.

Because Renaissance naturalists accumulated knowledge of an increasing number of kinds of plants and animals, the problem of keeping this knowledge organized in some meaningful way became increasingly urgent, as did the need for developing a coherent nomenclature for the large number of plants and animals they described. Their books and herbaria contained names in a virtual Babel of languages, and the names provided no clue as to how various organisms might be similar to each other.

Natural history, as pursued by both Renaissance humanists and medical writers, differed from the ancient discipline practiced by Aristotle and his disciple Theophrastus. Greek thinkers had viewed natural history as a preparation for natural philosophy. For them, natural history provided descriptions of plants and animals, for which natural philosophy would then provide explanations. In contrast to these ancient writers, practitioners of natural history during the Renaissance sought accurate descriptions of living things without concern for incorporating them into an all-embracing natural philosophy. The fact that the herbalists based their investigations on Dioscorides' book rather than Theophrastus'—whose works waited another century before being taken into account by Renaissance naturalists—demonstrates their emphasis on accurate description and medical application as opposed to philosophical explanation. In this way, they contributed to the formation of a new, autonomous discipline, natural history.

An examination of natural history in the Renaissance reveals intricate relationships among individual scholars, modes of communicating knowledge and specimens, and technological developments—notably printing. Commercial interests, such as the preparation and certification of drugs, provided some of the motives for establishing botanical gardens associated with faculties of medicine. A desire for descriptions of the wealth and exotic character of new colonial possessions motivated some monarchs to underwrite the activities of naturalists on exploratory expeditions to the New World. And competition among wealthy collectors led them to create what they called cabinets of curiosities, often the precursors to more rationally organized museums.

The work of Francis Bacon (1561–1626) increased the status and influence of natural history. Bacon, trained as a lawyer and jurist, was a philosopher and served as lord chancellor of England, a post he eventually lost after being charged with taking bribes. He wrote extensively on philosophy, which he sought to reform by stressing the utility of knowledge for improving the human condition and for compensating for the loss of knowledge humankind suffered as a consequence of Adam's fall. His means of accomplishing these lofty goals involved a total reform of philosophy, one in which natural history was to play a central role. Criticizing Aristotle's methods and abandoning the literary and philological methods of the Renaissance humanists, Bacon advocated an inductive approach to knowledge of the world. As the first step in this scheme he called for collecting as much factual information as possible.

Then, by gradually drawing generalizations from these facts, the natural historian would be able to proceed to increasingly broad and increasingly certain knowledge of the world. The most advanced stage of generalization would lead to certain knowledge of the real nature of things.

Bacon never completed his massive project to describe this new approach to knowledge and produce examples of natural histories. But his books inspired subsequent generations to pursue natural history. His utopian tale *The New Atlantis* (1627) describes a community devoted to collective research following his methods. The founding of the Royal Society of London in 1660 drew inspiration from Bacon's vision and incorporated the pursuit of histories of trades along with various natural histories as part of its mandate. The Baconians of the early Royal Society sought to catalog every aspect of the world—human as well as natural—in such histories.

In the years between Bacon's death and the founding of the Royal Society in 1662, most naturalists continued to pursue natural history in the Renaissance manner. For example, Edward Topsell (1572–1625), an Anglican clergyman, wrote *The Historie of Foure-Footed Beastes* (1607) and *The History of Serpents* (1608), drawing heavily from Gessner's five-volume *Historia animalium*. Like Gessner, Topsell listed the animals in alphabetical order and included accounts of mythological creatures as well as real ones. He endowed the animals he described with human and moral qualities. Topsell's purpose was unambiguously religious: he thought that knowledge of animals would lead to a greater appreciation of the power of the deity.

Even though descriptive natural history did not change much during the first half of the seventeenth century, debates about things dug out of the ground produced both heat and light. The Jesuit natural philosopher Athanasius Kircher (1602–80) argued that a "lapidifying [stony-making] virtue" that is distributed throughout the cosmos produces the stony character of "fossil objects" that resemble living things. A "plastic spirit" causes the particular shapes of these objects. Other naturalists rejected Kircher's animistic philosophy of nature and explained the formation of these objects very differently.

Niels Stensen (1638–86), better known by his Latinized name, Steno, proposed that fossils are the remains of living creatures that had became trapped in strata by the accumulation of sediment on the ocean floor. As a deeply religious individual, like most of his contemporaries (he was a convert from Protestantism to Catholicism, for which he actively proselytized and eventually became a bishop), he assumed a short biblical time scale of about 6,000 years for the age of the earth and consequently found the implications of his theory about fossils troubling. For Steno and many others in the seventeenth

## The Beaver

In *The Historie of Foure-Footed Beastes*
(1607), Edward Topsell reproduced the
images from Gessener's *Historia animalium*
and paraphrased Gesner's descriptions of
the animals. In the text accompanying this
emblematic image of the beaver, Topsell
refutes the ancient folklore about the beaver biting off its own testicles to keep the
hunters from catching him:

Teaching by the example of a Beaver, to give
our purse to theeves, rather than our lives, and
by our wealth to redeem our danger, for by this
means the Beaver often escapeth. There have
been many of them found that wanted stones
[testicles], which gave some strength to this
errour, but this was exploded in ancient time
for a fable; and in this and all other honest
discourses of any part of Philosophy, the only
mark whereas every good student and professor ought to aime, must be verity and not tales;
wherein the ancient have greatly offended . . . ;
and this poyson hath also crept into and corrupted the whole body of Religion. The Aegyp-
tians . . . when they will signifie a man that
hurteth himself, they picture a Beaver biting
off his own stones. But this is most false . . .

First, because their stones are very small, and so
placed in their body as are a Boars, and therefore impossible for them to touch or come by
them. Secondly, they cleave so fast unto their
back, that they cannot be taken away but the
beast must of necessity lose his life; and therefore ridiculous is their relation, who likewise
affirm, that when it is hunted (having formerly
bitten off its stones) that he standeth upright
and sheweth the hunters that he hath none for
them, and therefore his death cannot profit
them, by means whereof they are averted and
seek for another.

Topsell continued by discussing the uses of
the various parts of the beaver, such as the
skin and bones.

■ Edward Topsell, *The Historie of Foure-
Footed Beastes* (London, 1607).

century, the locations in which some fossils were found—such as sea shells embedded in strata high on mountains—also proved challenging. One common
explanation of these troubling facts was to say that the waters of Noah's flood
had deposited these shells in their strange locations.

Steno's theory of the organic origin of fossils received a warm reception
from Robert Hooke and other members of the Royal Society, in large part
because it seemed to illustrate the power of the corpuscularian theory of matter. As the organic origin of fossils became more widely accepted, Hooke and
others who presupposed a biblical account of the Creation faced two disturb-
ing facts. Some fossils appeared to be different from any known existing creatures. This raised the possibility of extinction, a possibility at odds with the

widespread assumption that all existing species had continued unchanged since the Creation. And the location of the fossils of sea animals embedded in mountainous strata raised questions about the history of the earth, questions that would perplex naturalists and philosophers for the next 150 years.

Among the fossils that provoked controversy because they resemble no known existing organism, ammonites played a significant role. Ammonites are the fossil remains of an extinct form of cephalopod related to the modern octopus, squid, and nautilus. Because they resembled no extant creatures, they became the focus of controversy in early modern natural history. Although Martin Lister (1639–1712), a physician and naturalist, was willing to accept Steno's claim that these fossils from Italy had organic origins, he rejected the idea that ammonites were the remains of animals because he could find no trace of shell in the fossils. Also the fact that different sorts of fossils were embedded in different kinds of rock seemed to refute Hooke's idea that the sea had tossed up the shells, because in that case the same kinds of fossils should be found everywhere. In debates with Lister, Hooke was willing to acknowledge that ammonites had become extinct, but he thought that every extinct creature was replaced by another, thus preserving the fullness of God's creation.

The nature and origin of ammonites and other fossils remained controversial for decades. Lest the speculative tenor of the controversy be laid only at the doorstep of thinkers embracing natural theology, we should note that the paragon of Enlightenment *philosophes*, François Marie Arouet (1694–1778)— better known by his pen name, Voltaire—rejected the organic origin of fossils because they pointed to profound changes in the history of the earth, something he considered impossible in a stable Newtonian universe. Accordingly, he thought that the shells found on mountaintops had fallen off the hats of crusaders as they returned from defending the Holy Land and that ammonites are petrified snakes.

John Ray (1627–1705), the most prominent naturalist in England during the second half of the seventeenth century, wrote extensively on the flora of the British Isles, as well as on birds and fishes. His work reflected traditional approaches to natural history but also rested on careful observations of the specimens he described. His strong background in classical languages and Hebrew enabled him to engage in the sophisticated philological practices that had characterized natural histories written by his predecessors. Considerations of the names of plants and animals, however, were only the beginning of his descriptions, which he based on careful observation. He identified the locations in which he found his specimens, and he described them in meticulous detail.

Ray did not develop a complete scheme for classifying plants, but he did describe a method for arranging them into general categories such as trees, shrubs, and herbs. He subdivided these categories based on such criteria as the kinds of fruit, the kinds of flowers, or the kinds of seeds the plants produced. In various books he listed plants in alphabetical order. A number of naturalists did try to arrange organisms into rational schemes, but problems of classification and nomenclature were not sorted out until Carolus Linnaeus (1707–78) established the foundations of modern classification in *Systema naturae* (*System of Nature*), which went through numerous editions during his lifetime.

For Ray, religious significance permeated natural history. The biblical story of the Creation formed the backdrop for his studies. Natural theology provided the foundation for his view that all the kinds of plants and animals that now exist have existed ever since the Creation and that, because God rested on the seventh day, there can be no change in the pattern of the creation. He did recognize the existence of some anomalies. He once planted some cauliflower seeds and noted that some of the seeds produced only cabbages. He explained this apparent change as an example of degeneration, comparable to Adam's fall in the Garden of Eden. He ascribed the troubling mountaintop location of fossils to the results of Noah's flood. Ray considered natural history to be essentially an act of worship. In an immensely influential book, *The Wisdom of God Manifested in the Works of the Creation* (1691), he argued that the study of the natural world reveals God's goodness, power, and wisdom. Many other naturalists and natural philosophers at the time held similar views, and natural theology—seeking knowledge of God through study of the natural world—was a prominent feature of late-seventeenth-century thought.

## Challenging Galen: Anatomy and Physiology

Natural history, a fundamentally descriptive endeavor, left unanswered other questions about the functioning of bodies, particularly human bodies. Medical scholars carried out the study of these questions usually in the context of faculties of medicine. Challenges to Galen's account of the cardiovascular system followed in the wake of Vesalius' work in anatomy. Realdo Colombo (1510–59), who had worked as Vesalius' assistant, succeeded Vesalius to the chair of surgery and anatomy at the University of Padua. Although Colombo retained a basically Galenic outlook, he modified the account of the cardiovascular system in one important way by arguing for the pulmonary circulation. This idea, first proposed by Ibn al-Nafis (1213–88), who came from Damascus and became chief of physicians in Egypt, probably arrived in Italy in the course of trade with the eastern Mediterranean.

According to Colombo, blood passes from the venous system through the lungs and then through the heart into the arterial system. He argued that blood from the venous system passes through the pulmonary artery to the lungs, where it is thinned. The pulmonary vein then carries this thinned blood into the left atrium and from there into the left ventricle. He observed that in dead bodies the pulmonary vein is always full of blood, not spirit. He argued that the valves in the heart ensure the one-way flow of blood, thus supporting the idea of pulmonary circulation: when the heart contracts, the valves prevent the blood from flowing back along the same path from which it came. He argued further that the heart cannot create new blood, and he thus ruled out ebbing and flowing, replacing that mechanism in part with pulmonary circulation.

In the next generation of Paduan medical thinkers, Fabricius of Aquapendente (1537–1619) continued the anatomical work started by his predecessors. Like Colombo, he held a fundamentally Galenic understanding of physiology, but he made anatomical discoveries that would eventually undermine that tradition. Fabricius' most dramatic discovery was that the veins contain valves that open only towards the heart, ensuring that venous blood always moves toward the heart. He argued that the one-way flow of blood in the veins prevents blood from accumulating in the lower body, an eventuality that would cause a permanent distension of the hands and feet and would cause the upper parts of the body to be undernourished. He claimed that there are no valves in the arteries because the constant ebbing and flowing of arterial blood eliminates the dangers that would occur in the veins if there were no valves. Despite his discovery of the valves in the veins, a discovery that would seem to undermine Galen's account of the cardiovascular system, Fabricius did not draw the apparently obvious conclusion.

That challenge came from one of Fabricius' students at Padua, the Englishman William Harvey (1578–1659). Harvey, who became the personal physician of the English king Charles I, took the final, dramatic step in refuting Galen's account of the cardiovascular system. Fully absorbing the recent discoveries and using careful observations and experiments, Harvey proved that the blood circulates through the body. He published an account of his methods and results in *Exercitatio anatomica de motu cordis et sanguinis in animalibus* (*Anatomical Dissertation on the Motion of the Heart and Blood in Animals*) (1628).

Harvey opened his treatise by demonstrating that Galen's account of the vascular system cannot be true. He accepted the fact of pulmonary circulation, adding some strong arguments to those already established. Structural similarity between the right and left ventricles implies that they have similar functions. He also observed that the pulmonary vein and pulmonary artery are always full of blood, not air, spirit, or sooty wastes, as Galen had claimed. And close examination of the septum revealed no pores. He concluded that the evidence for pulmonary circulation was compelling.

Harvey then faced the question of how the blood moves through the rest of the body. Observing the motions of the heart, he stated that the heart acts when it contracts, at which time it issues a spurt of blood. Measuring the quantity of blood in a spurt and multiplying that quantity by the number of times the heart beats in a minute, Harvey stated that digested food cannot possibly produce the quantity of blood that flows through the heart. For this reason he proposed that the same quantity of blood constantly circulates through the body. He performed some simple experiments with living animals to establish the one-way flow of this blood. When he tied off the vena cava with a ligature, the part of the vena cava between the ligature and the heart quickly became empty of blood; and, if he tied off the aorta, the portion between the ligature and the heart became distended with blood. He then performed a series of experiments, in which he tied off blood vessels in the limbs. When he used a very tight ligature to tie off the arteries, blood backed up closer to the heart, but the extremities became cold. It followed that the heart pumps blood into the arteries, through which blood travels towards the extremities. When he used a slightly looser ligature to tie off the veins—which lie closer to the surface—but not the arteries, the extremities remained full of blood, but the veins between the ligature and the heart were emptied of blood.

The fact that the valves in the veins allow only one-way flow towards the heart, coupled with the facts revealed by the experiments with ligatures, led him to conclude that the blood flowing through the veins incessantly returns to the heart. Thus, the blood constantly flows through the body in a figure eight: the heart pumps blood into the aorta, from which the blood travels into the smaller vessels of the arterial system; it passes into the veins, returns to the heart via the vena cava; then the right ventricle pumps the blood into the pulmonary artery, from which it passes through the lungs and then back into the heart via the pulmonary vein and into the left ventricle, from which it is once again pumped into the aorta.

A troubling question remained: How does the blood from the arteries pass into the veins? Like Galen, Harvey needed to postulate some connection between the two parts of the vascular system. He hypothesized that there exist tiny vessels—which he called *anastomoses* (the same word used by Galen for the supposed pores in the septum)—at the ends of the arteries through which the

## William Harvey's Experiments with Ligatures

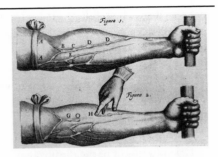

In Figure 1 the ligature ties off the veins in the arm. *A* through *E* mark places where the valves in the veins prevent the blood from flowing back down the arm into the hand. In Figure 2 Harvey uses a finger to block the upward flow of the blood; this distends the vein above the valve and prevents the blood from flowing back down the arm. In Figure 3 Harvey demonstrates that if you stroke the vein at *H*, causing the blood to flow upward through that valve, and then depress the vein at *O*, the blood above *O* will become empty of blood, which continues to flow towards the heart through the veins above *O*. In Figure 4 similar results occur. Harvey concluded that the valves in the veins ensure one-way flow of venous blood towards the heart.

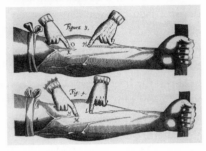

■ William Harvey, *De motu cordis* (London, 1628).

blood passes from the arterial system into the venous system. Harvey could not observe these *anastomoses*. In the next generation, Marcello Malpighi (1628–94), using the newly invented microscope, discovered the capillaries in the tissue of the lungs, thus confirming Harvey's hypothesis of a direct connection between the arteries and veins.

In establishing the circulation of the blood, Harvey significantly revised Galenic physiology and rejected Galen's conclusions if not of his methods. But the new understanding of the cardiovascular system raised many new questions. For example, Why does the blood flow through the lungs? In the 1660s several English natural philosophers, members of the early Royal Society, performed experiments addressing this question. Experimenting with the air pump, Robert Boyle established that air is necessary for maintaining the life of animals. He placed small animals into the receiver of his pump. When he pumped out the air, the animals died. Using live dogs, Richard Lower (1631–91) demonstrated that the passage of air through the lungs, not the motions associated with breathing, kept the dogs alive. The experiment involved keeping the dog's chest full of air by using a bellows to pump air through the lungs. As long as air continued to flow through the lungs, even though the lungs did not expand and contract, the dog stayed alive. Further experiments demonstrated that exposure to the air causes arterial blood to become bright red, in contrast to the darker color of venous blood. These observations led to a variety of speculations about what component of air supports life and causes the blood to change color.

Although the work of Harvey and his followers led to a complete revision of Galen's account of the cardiovascular system, their conclusions were not the decisive factor in the demise of Galenic physiology. Rather, medical thinkers who had adopted the mechanical philosophy rejected Galen's main causal mechanism, attraction, as a kind of action-at-a-distance that they sought to eliminate from natural philosophy. Similarly, Galen's claim that the major organs—the liver, heart, and brain—possess active faculties, each responsible for different types of spirits, did not fit into the new, mechanical assumptions about the workings of nature.

Careful observation, enhanced by experiments, characterized the research conducted by these medical men. If Vesalius had begun his work in anatomy in order to correct Galen's mistakes, his work precipitated the downfall of Galen's account of the cardiovascular system. There is some irony in these developments, as Galen himself had been a keen observer, and his errors in human anatomy had resulted at least in part from the lack of human cadavers. Moreover, these seventeenth-century medical thinkers retained Galen's teleological assumption that form follows function.

## Animating Matter: The Souls of Animals and Humans

The study of living things exacerbated a question that had concerned philosophers at least since the time of Aristotle: What distinguishes humans from animals? Christian thinkers claimed that only humans possess an immortal soul. The immortality of the human soul remained a focus of concern through the Renaissance, particularly in reaction to the Aristotelian philosopher Pietro Pompanazzi (1462–1525), who maintained that the human soul is material and perishable. The question became increasingly urgent as the rise of the mechanical philosophy raised the specter of materialism, the view that only matter exists and that everything in the world—including the human soul—is material. Aware of this danger, virtually all the mechanical philosophers argued for the immateriality and immortality of the human soul. They employed the traditional tactic of comparing the human soul with the souls of animals in order to determine what characteristics were uniquely human. The mechanical philosophers used the immaterial, immortal soul as a way to define the limits of mechanization. They did not think that mechanical explanations encompassed everything in the world.

Like Aristotle, whose *De anima* counted as one of his natural books, both Gassendi and Descartes included discussions of the soul in their books on natural philosophy, an important indication that some of the Aristotelian disciplinary categories lingered, even within the context of the new mechanical philosophy. Gassendi claimed that humans possess two souls, a material soul and an immaterial, immortal soul. The material soul, which animals also possess, is composed of extremely tiny, very active atoms. Diffused through every part of the body, it produces the organism's vitality. At the time of death it dissipates into the air like smoke. Humans alone possess immaterial, immortal souls. Gassendi argued for the existence of this uniquely human soul by noting, among other things, that humans, unlike animals, are able to engage in self-reflective, abstract thinking. Animals, he said, can recognize universals, as when a dog can discern whether an approaching animal is another dog or a human. But dogs, unlike humans, cannot contemplate the nature of universality. This kind of self-reflection, Gassendi argued, can only be an attribute of something immaterial. Deploying an argument commonly used at the time, he claimed that immortality follows from immateriality because something immaterial cannot be divided and therefore cannot be destroyed.

Descartes also claimed that there is a fundamental difference between humans and animals. His famous, foundational argument, "I think, therefore I am," ostensibly established the existence of an immaterial mind or soul, the thinking agent. Animals, he thought, were simply material automata. They do not possess cognition or even sensation. The motions that a dog exhibits when kicked—motions that appear to indicate that the dog feels pain—are simply reflex actions caused by the motion of fluids in the dog's nerves. Unlike humans, dogs actually lack the ability even to feel pain, never mind to think.

Boyle devoted many treatises to proving the existence of an intelligent, beneficent God who imparts his purposes into the universe. As part of his effort to avoid the dangers of materialism, Boyle claimed that in the course of the development of every human life, God infuses an immaterial, immortal soul into the fetus. God does not miraculously intervene in this manner in the lives of animals.

For these and many other mechanical philosophers, the difference between humans and animals served to mark the boundary between the material world and the spiritual world. In striking contrast to these seventeenth-century thinkers, philosophers today attempt to define humans in contrast to machines—computers. This shift surely marks a decline in human self-esteem.

Many different strands formed the fabric of the study of living things in early modern Europe. Medical botany, explorations of the New World, the humanist revival of ancient texts, and the development of observational and experimental methods all played important roles. The pursuit of knowledge of living things bore on the preoccupations of the time, notably medicine and theology. Although these studies gained status and importance, the various activities did not become united under a single disciplinary rubric. The word "biology" and the idea that it signifies—a study of living things in general—did not emerge until sometime in the nineteenth century.

# 8 Rethinking the Universe

## *Newton on Gravity and God*

Many developments in seventeenth-century natural philosophy, mathematics, and astronomy culminated in the work of Isaac Newton. In his major published works in natural philosophy, the *Philosophiae naturalis principia mathematica* (*Mathematical Principles of Natural Philosophy*) (1687) and the *Opticks* (1704), Newton developed new methods for solving problems in physics. His stunning contributions to physics, cosmology, and mathematics served as models and set the research agenda for natural philosophers in generations to come. Not only did his physics provide answers to questions that had remained outstanding since the time of Copernicus, but his work also embodied a major shift in disciplinary boundaries. The concept of force, central to his work in natural philosophy, enabled him to achieve the goal sought by the mechanical philosophers—to explain natural phenomena in terms of matter and motion. At the same time, his use of the concept of force introduced fundamental changes into that philosophy of nature.

Newton's interests were broad, extending far beyond his work on motion, light, and mathematics. He spent many years absorbed in theological and alchemical studies. Although he published little or nothing in these subjects, his voluminous manuscripts provide irrefutable evidence of his abiding concern with these matters. At the time of his death, his papers and manuscripts went for safekeeping to Catherine Barton Conduitt, his half-niece, who had married John Conduitt, one of his earliest biographers. The physician Thomas Pellet (1689–1744)—later president of the Royal College of Physicians—and a number of other individuals close to Newton examined the papers to determine which ones were suitable for publication. Other than immediately publishing *The Chronology of Ancient Kingdoms, Amended* (1728), a work on ancient empires aimed at demonstrating the priority of Hebraic civilization, Pellet labelled the bulk of the manuscripts—particularly those on alchemy and those revealing Newton's heretical theology—"Not fit to be printed." They ultimately passed into the hands of his relatives, where they languished for almost two hundred years. Eventually, in the 1870s, the family gave all papers to the University of Cambridge, where they are now known as the Portsmouth Collection. The university, however, returned the alchemical and theological papers to the family.

In 1936 the family sold the remaining papers—those "not fit for to be printed," which included most of Newton's theological and alchemical manuscripts—in an auction handled by Sotheby's in London. Although the sale dispersed the papers widely, John Maynard Keynes managed to retrieve a third to a half of them and left them to the library of King's College Cambridge. Abraham S. Yahuda, a Palestinian Jew and major Arabic scholar, acquired most of the theological papers at the Sotheby's auction. Having been unsuccessful in convincing several university libraries in the United States to take the manuscripts, he decided to leave his collection to the National Library in Jerusalem. Consequently, only after the middle of the twentieth century did scholars have access to the full range of Newton's thought.

## Mathematizing Natural Philosophy: The Concept of Universal Gravitation

Access to Newton's manuscripts has enabled historians to trace the development of his thinking to a depth not possible by considering his published works alone. Newton's fame stems largely from his mathematics, his optics, and his theory of universal gravitation, which answered two remaining problems generated by Copernican astronomy: What holds the planets in their orbits? and, Can there be a physics that applies to both the heavens and the earth? Newton's answers to these questions resulted from work that he started during his undergraduate studies, when, among other things, he considered problems about matter, gravity, and impact. Aware of the work on motion by his predecessors Galileo, Descartes, and Huygens, he came to understand that a science of motion built on the principle of inertia requires a concept of force.

In 1666, when he had returned to his twice-widowed mother's farm during an outbreak of plague that forced the university at Cambridge to close, he asked himself whether the gravity that causes heavy bodies close to the surface of the earth to fall—like an apple from a tree—extends all the way to the moon. Using his own analysis of circular motion, he realized that for the moon to remain in a stable orbit there must be a force pulling it towards the earth precisely equal and opposite to the centrifugal force, which would cause it to fly off into space. Newton coined the term "centripetal force" to designate this attractive force. He hypothesized that the force of gravity varies inversely as the square of the distance between the earth and moon. His knowledge of Kepler's laws, particularly the third law, played a central role in his calculations. However, faulty data for the size of the lunar orbit caused his calculation to fail, and so he also failed to prove his hypothesis about the force of gravity. At this point he set aside the calculation. In his early manuscripts on motion, Newton used the concept of force simply as a mathematical expression for measuring a body's deviation from inertial motion and did not address the deep problems it posed for the mechanical philosophy.

During his absence from Cambridge, which lasted for about eighteen months, Newton also worked up his ideas about light and colors. Newton's accomplishments in physics depended on his invention of the calculus, which he called his "method of fluxions." He achieved his major mathematical insights during this interlude as well but did not publish them at the time. His method of fluxions played a critical role in the demonstrations of propositions in the *Principia*, although he camouflaged the new methods behind a smokescreen of Euclidean geometry. Among other things, his theory of fluxions enabled him to derive the forces causing orbital motion and to calculate the areas called for by Kepler's second law. No wonder historians have labeled this period Newton's *annus mirabilis* (marvelous year).

Following eighteen months on his mother's farm, Newton returned to Cambridge, where he had won a fellowship at Trinity College. In the late 1660s, he wrote up his experiments on light and colors and eventually sent the paper to Henry Oldenburg (1619–77), corresponding secretary of the Royal Society, who ensured that they were published in the *Philosophical Transactions of the Royal Society* in 1672. The controversy that greeted his paper soured Newton on presenting his work publicly. He found Robert Hooke's rejection of the analysis of white light and Hooke's resurrection of the traditional theory that colors are produced by the modification of white light particularly irritating. For about the next ten years, Newton did not communicate with the Royal Society. During that period he devoted himself to mathematics, theology, and alchemy.

When Hooke became corresponding secretary of the Royal Society following Henry Oldenburg's death, he wrote to Newton in an attempt to restore communications between Newton and the Royal Society. The two men corresponded briefly about problems concerning the motions of orbiting bodies. In one letter Hooke proposed that the central force attracting an orbiting body would be inversely proportional to the square of the distance between them, but he offered no justification for that suggestion. After the correspondence broke off, Newton demonstrated that an elliptical orbit around an attracting body at one focus entails an attraction that is inversely proportional to the square of the distance between the bodies. At the time, however, he did not publish the inverse-square law. Hooke later accused Newton of plagiarizing this result, but this accusation cannot be sustained in light of Hooke's inadequate knowledge of mathematics.

A visit from the young astronomer Edmond Halley (1656–1742) in 1684 propelled Newton to work out his ideas on orbital motion in mathematical detail. One of his early biographers reported:

> In 1684, D$^r$ Halley came to visit him at Cambridge, after they had been some time together, the D$^r$ asked him what he thought the Curve would be that would be described by the Planets supposing the force of attraction towards the Sun to be reciprocal to the square of their distance from it. S$^r$ Isaac replied immediately that it would be an Ellipsis, the Doctor struck with joy & amazement asked him how he knew it, what saith he I have calculated it, whereupon D$^r$ Halley asked him for his calculation without any farther delay, S$^r$ Isaac looked among his papers but could not find it, but he promised him to renew it, & then send it to him.[1]

Newton's attempt to recover those calculations and to place them on a secure foundation led to his writing the *Principia*, a book which was the culmination of developments in astronomy and physics since the time of Copernicus. This achievement rested on the concept of force. "The basic problem of [natural] philosophy," Newton wrote, "seems to be to discover the forces of nature from the phenomena of motions and then to demonstrate the other phenomena from these forces."[2] The organization of the *Principia* reflects this dictum. In Books I and II Newton developed the mathematical characteristics of several different attractive forces, with particular attention to a force that varies inversely as the square of the distance between mutually attractive bodies. In Book III, he used the fact that the planets obey Kepler's laws to establish that gravitational attraction is indeed such a force. He then justified his claim that there is a force of universal gravitation affecting all bodies in the universe by demonstrating that such a force of universal gravitation can explain several phenomena that other theories had failed to explain.

Newton began Book I of the *Principia* by enunciating three laws of motion:

1. Abraham de Moivre, "Memorandum of 1727," in *Early Biographies of Isaac Newton: 1660–1885*, ed. Rob Iliffe, Milo Keynes, and Rebekah Higgitt, 2 vols. (London: Pickering, 2006), 1:124–25.

2. Isaac Newton, *The Principia*, trans. I. Bernard Cohen and Anne Whitman, assisted by Julia Budenz (Berkeley and Los Angeles: University of California Press, 1999), p. 382.

Law 1. Every body perseveres in its state of being at rest or moving uniformly straight forward, except insofar as it is compelled to change its state by forces impressed.

Law 2. A change in motion is proportional to the motive force impressed and takes place along the straight line in which that force is impressed.

Law 3. To any action there is always an opposite and equal reaction; in other words, the actions of two bodies upon each other are always equal and always opposite in direction.[3]

Together these three laws defined what Newton meant by "force" in the *Principia* and provided the basis for all the demonstrations that followed.

Newton devoted most of Book I of the *Principia* to the mathematical exploration of various attractive forces. He proved that if a body orbits around a central force in a path that is a conic section, the force must be the inverse of the square of the distance between the source of the central force and the orbiting body. Conversely, if there is a centripetal force that varies as the inverse of the square of the distances between two bodies, an orbiting body will follow the path of one of the conic sections. In other words, Kepler's laws, according to which the planets move in elliptical orbits around the sun at one focus, entail the existence of a central force between the sun and the planets that obeys the inverse-square law.

For the proofs in Book I of the *Principia* Newton used concepts such as "vanishing quantities" and sums taken to infinity that indicate that he actually used the theory of fluxions (i.e., calculus) in his reasoning. However, because this new branch of mathematics lacked foundations and seemed counterintuitive, he presented his reasoning in the ancient, traditional form of geometrical proofs. For Newton, who believed that he was restoring an ancient wisdom (called the *prisca sapientia*), the ancient methods were superior to those of his contemporaries. Other natural philosophers and some mathematicians criticized his theory of fluxions for allowing such seeming impossibilities as having something become nothing and for claiming to find a finite sum for an infinite series. As confidence in the new mathematics grew, later generations of Newtonians translated his mathematics into an algebraic and analytic form, which explicitly used the calculus. Although the calculus proved to be an extremely useful mathematical tool and mathematicians developed its methods extensively in the eighteenth century, rigorous foundations for this impor-

3. Ibid., pp. 416–17.

## Diagram from Newton's Proof of the Inverse-Square Law

For the inverse-square law, Newton's method of proof consisted of considering the forces acting on an orbiting body at individual points on the orbit. At each point—*A* through *F*—he analyzed the forces that caused the body to deviate from inertial motion in a straight line tangential to the orbit. To determine the precise shape of the orbit, he considered the intervals between those points to become smaller and smaller until they vanished altogether. The resulting calculation established that an inverse-square force produces an elliptical orbit.

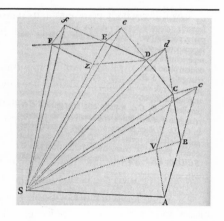

■ Isaac Newton, *Philosophiae principia mathematica naturalis* (London: The Royal Society, 1687), p. 37.

tant branch of mathematics were not established until the late nineteenth and early twentieth centuries.

In addition to orbital motion, Newton also demonstrated that, given the inverse-square law, bodies close to the surface of a large body like the earth will conform to Galileo's law of falling bodies. In other words, Newton proved that the same force of gravitation governs the motions of the planets and the behavior of bodies close to the earth's surface. Here he took the final step in moving away from Aristotelian cosmology: he showed that the same laws of physics apply to the terrestrial and celestial regions, finally eliminating that distinction and establishing the spatial uniformity of nature.

Continuing his mathematical analysis in Book II of the *Principia*, Newton examined the motion of bodies through a fluid medium. He demonstrated that, because of the medium's resistance, an orbiting body in a fluid medium would quickly slow down and stop. This demonstration refuted Descartes' theory of vortices, the only other mechanism that natural philosophers had proposed to explain the motions of the planets. "The hypothesis of vortices," Newton declared, "is beset with many difficulties."[4]

In Book III, Newton added empirical data to his mathematical analysis. From the fact that the planets, the sun, the moon, and the satellites of Jupiter

4. Ibid., p. 939.

all obey Kepler's laws, he demonstrated that there must exist a force of mutual attraction that varies inversely as the square of the distance between each planet and the sun and each satellite and the planet around which it revolves. Extrapolating from the solar system, he claimed that there exists a force of universal gravitation such that every body in the universe attracts every other body by a force that is measured as the inverse of the square of the distance between them and the product of their masses, as expressed by

$$F \propto \frac{Mm}{r^2} \, ,$$

where $M$ and $m$ are the masses of the two bodies and $r$ is the distance between them. The law of universal gravitation answered the two outstanding questions raised by heliocentric astronomy: What holds the planets in their orbits? And how can celestial and terrestrial physics be united? By answering these questions successfully, Newton's theory of universal gravitation completed the revolution that Copernicus had begun.

After demonstrating the existence of the force of gravitation as governing the relationships between the sun and the planets, Newton used this force to explain four hitherto unexplained phenomena: the ebb and flow of the tides, the variations of lunar motions, the precession of the equinoxes, and the motions of comets. These phenomena had hitherto been difficult if not impossible to explain, although some of the best minds of the century had addressed them. By demonstrating that the theory could explain these phenomena—phenomena that had not played a role in formulating the theory of universal gravitation—Newton further justified his theory. The theory of universal gravitation gave future generations of physicists and astronomers a powerful tool for solving problems in physics and cosmology.

The *Principia* is significant not only because of Newton's stunning contribution to physics and cosmology, but also because of its impact on the classification of disciplines. Remember that in the traditional Aristotelian classification, mathematics and physics were separate disciplines, based on different principles and following different methods. In the Aristotelian context, Newton's title—*The Mathematical Principles of Natural Philosophy*—sounds like an oxymoron. Rather than expressing a contradiction, however, the title is a bold statement of the changes in disciplinary boundaries that had occurred over the previous 150 years. For the next two centuries, the methods and structure of the *Principia* served as a model that natural philosophers in many fields emulated.

## Explaining the Phenomena: Matter and the Problem of Attraction

Newton's remarkable accomplishments and profound rearrangement of the disciplines rested on his concept of force. But the concept of force seemed to contradict one of the basic tenets of the mechanical philosophy, namely, the elimination of all action-at-a-distance in the world. What led Newton to articulate such a concept? And how did he address the problems it raised?

As an undergraduate at Trinity College Cambridge in the early 1660s, Newton read widely in the major texts of seventeenth-century natural philosophy. A notebook that he kept during this period reveals his knowledge of the writings of Descartes, Gassendi, Hobbes, Boyle, and other natural philosophers of the time. Newton rejected the Aristotelianism that was still being taught in the university's official curriculum and endorsed the new mechanical philosophy. He assumed that all natural phenomena can be explained in terms of matter and motion alone, and he set out to determine which version of the mechanical philosophy—Descartes' or Gassendi's—was better. In the first section of the notebook, entitled "Of the First Matter," Newton considered their respective theories of matter and opted for an atomism similar to Gassendi's. He proceeded to discuss a wide range of topics—motion, place, density, colors, the tides, gravity, magnetism, electricity, and many others—all of them requiring explanation within any natural philosophy. In each case, with his keen physical intuition and his mind filled with the mechanical view of nature, Newton designed thought experiments to put various mechanical explanations to potentially empirical test.

For example, he considered mechanical explanations of gravity, a departure from the Aristotelian tradition that had regarded gravity as an innate property of heavy bodies. Descartes had attempted to explain gravity by claiming that in the vortices some bodies have less centrifugal tendency than the subtle matter that he thought fills all space, and so in the vortex around the earth, bodies will tend to remain closer to the center and thus will appear to be heavy. Gassendi had explained gravity in terms of hooks attached to lines of magnetic attraction, envisaging the lines as physical. Both explanations used only mechanical terms. In this case, Newton rejected both explanations, the one because he did not agree that all of space is full of subtle matter and the other because it implied that gravity is proportional to the surface area of a body when in fact it is proportional to the quantity of matter in the body.

For a period of about fifteen years following his undergraduate studies, Newton attempted to develop a philosophy of nature that remained true to the demands of the mechanical philosophy: all phenomena are caused by matter and motion; matter is inert; and action-at-a-distance does not exist. Certain phenomena, however, seemed to resist explanation in purely mechanical terms, and Newton had already been thinking about them in his student notebook. Gravity, magnetism, the reflection and refraction of light, capillary action, surface tension, the expansion and contraction of air, the cohesion of bodies, and certain chemical phenomena occupied him in the early 1660s and continued to trouble him throughout his life. The resistance of these phenomena to mechanical explanation ultimately contributed to his radical revision of the mechanical philosophy.

In the mid-1670s, Newton wrote two papers in which he attempted to explain various phenomena in terms of strictly mechanical principles. He constructed speculative, invisible mechanisms to explain the appearances by appealing to an aethereal medium composed of particles of the same kind of matter as air, but rarer, subtler, and more strongly elastic. Newton used this aether to explain electrical attraction, gravitation, the cohesion of bodies, sensory perception, animal motion, heat, and optical phenomena. For example, he explained gravitation by proposing that the descent of aether towards the earth carries bodies down. Since aether is most dense where ordinary matter is least dense, and vice versa, aether gradients exist between media of different optical densities. Such gradients cause the reflection and refraction of light as it passes from one medium to another. Aether gradients also explain the difficulty of pressing two pieces of glass together so that they touch. When two bodies are brought close to each other, the aether between them must be rarefied before they can touch, and the rarefaction of the aether requires the application of force. Because of the aether, bodies have an "endeavour to recede from one another."[5] For the same reason, flies can walk on water without wetting their feet, and a pile of dust does not cohere even when tightly compacted.

Bodies soluble in water dissolve because the particles of water enter the pores of the body and, by equalizing the pressure of the aether on all sides of a particle, tend to shake it loose. But not all bodies are soluble in water. Water cannot enter the pores of metal to dissolve it. The explanation is "not that water consists of too gross parts for this purpose, but because it is unsociable to metal. For there is a certain secret principle in nature by which liquors are sociable to some things & unsociable to others."[6] This "secret principle" of

5. Newton to Robert Boyle, 28 February 1679, in *The Correspondence of Robert Boyle*, ed. Michael Hunter, Antonio Clericuzio, and Lawrence M. Principe, 6 vols. (London: Pickering and Chatto, 2001), 5:144.

6. Ibid., p. 145.

sociability addressed the problem of specificity, which Newton had already noted: magnets attract iron but not copper; water sinks into wood but not metals; mercury sinks into metals but not wood; *aqua fortis* (nitric acid) dissolves silver but not gold; *aqua regia* (a mixture of hydrochloric acid and nitric acid) dissolves gold but not silver. This principle of sociability also allowed him to explain other difficult phenomena. Talk of sociability may have increased the power of his philosophy of nature, but it also marked a major departure from the orthodox mechanical philosophers who had insisted that matter is passive and cannot act at a distance. Endowing matter with sociability and unsociability presupposes that matter has some kind of innate activity, the very thing that the mechanical philosophers aimed to expunge from the natural world.

Newton's attempt to forge an aether with purely mechanical properties reached a crisis around 1679. In an incomplete manuscript entitled "De aere et aethere" ("On Air and Aether"), he began by describing the properties of air: "Among the properties of air, its great rarefaction and condensation are remarkable."[7] He ascribed these properties to the fact that the particles of air and other bodies repel each other at a distance. The pages that follow contain explanations of various physical and chemical phenomena in terms of these properties of air. The mutual repulsion of the particles of air is an instance of action-at-a-distance and demanded mechanical explanation. Newton intended to explain this repellency between particles of air and gross matter in terms of some aethereal mechanism.

He began the section on aether by asserting the existence of an aether, composed of particles of the same kind of matter as those of air: "And just as bodies of this Earth by breaking into small particles are converted into air, so these particles can be broken into lesser ones by some violent action and converted into yet more subtle air which, if it is subtle enough to penetrate the pores of glass, crystal and other terrestrial bodies, we may call the spirit of air, or the aether."[8] Newton then undertook the explanation of magnetism and static electricity in terms of this subtle, mechanical aether. The treatise ends abruptly in the middle of a sentence. Newton must have suddenly realized that he was embarking on an infinite regress. If aether is composed of smaller particles of the same matter as air, the particles of aether must also be mutually

7. Isaac Newton, "De aere et aethere," in *Unpublished Scientific Papers of Isaac Newton*, ed. and trans. A. Rupert Hall and Marie Boas Hall (Cambridge: Cambridge University Press, 1962), p. 221.
    8. Ibid.

repellant. How was this action-at-a-distance to be explained without invoking a micro-aether, and then a mini-micro-aether, and so on *ad infinitum*?

At this point and for about the next thirty years, Newton abandoned aethereal speculations. In addition to the problems he had encountered in formulating a coherent account of a mechanical aether, certain experimental results reinforced this decision. Experiments with a pendulum in an evacuated container demonstrated that there is actually so little resistance to the pendulum's motion that the aether would have to be so rare as to be useless for the purposes for which Newton intended it. Further, his studies of motions through resisting media produced evidence against the existence of aether in the region of the planets. For, as he proved in the *Principia*, resistance is proportional to the mass of the medium. Thus, no matter how finely divided the particles of an interplanetary aether might be, it would still retard the motion of the planets in a manner contradicted by the observed constancy of their motions.

As a result of these considerations, Newton replaced his aethereal speculations by adding forces of attraction and repulsion.

> Have not the small Particles of Bodies certain Powers, Virtues, or Forces, by which they act at a distance, not only upon the Rays of Light for reflecting, refracting, and inflecting them, but also upon one another for producing a great Part of the Phaenomena of Nature? For it's well known, that Bodies act one upon another by the Attractions of Gravity, Magnetism, and Electricity; and that these Instances show the Tenor and Course of Nature, and make it not improbable but that there may be more attractive Powers than these. For Nature is very consonant and conformable to her self.[9]

In a speculative "Query," first appended to the Latin edition of the *Opticks* in 1706, Newton explained that the passive matter of the traditional mechanical philosophy did not suffice to account for the phenomena; adequate explanations required some kind of active principle. Given the inertial property of matter, the quantity of motion in the world would diminish in the absence of an active principle that would serve as a continual source of motion.

> The *Vis inertiae* [force of inertia] is a passive Principle by which Bodies persist in their Motion or Rest, receive Motion in proportion to the Force impressing it, and resist as much as they are resisted. By this Principle alone there never could have been any Motion in the World. Some other Principle was necessary for putting

9. Isaac Newton, *Opticks; or, A Treatise of the Reflections, Refractions, Inflections & Colours of Light*, repr. from the 4th ed. (1730), with a foreword by Albert Einstein, introduction by Sir Edmund Whittaker, and preface by I. Bernard Cohen (New York: Dover, 1952), pp. 375–76.

Bodies into Motion; and now that they are in Motion, some other Principle is
necessary for conserving the Motion. . . . Motion is much more apt to be lost than
got, and is always upon the Decay.[10]

In the sense that he continued to explain natural phenomena by means of the
properties of matter, Newton remained true to the aims of the mechanical
philosophy. But with attractive and repulsive forces, he introduced action-at-
a-distance. He had now come to ascribe to matter the very properties that the
mechanical philosophers had deliberately banned.

Newton's growing acquaintance with chymical phenomena—he left at least
a million words in manuscript on alchemical subjects—reinforced his perception of specificity and activity in nature. The manuscripts reveal that Newton
was deeply immersed in alchemical studies. A central feature of alchemy is
that it describes a natural world that is full of activity. His alchemical studies
probably contributed to his willingness to add attractive and repulsive forces
to the conventionally mechanical properties of bodies.

The forces that Newton introduced after the 1670s replaced one-for-one
the earlier explanations that he had couched in terms of aether gradients. He
remained preoccupied with elective affinities, differences in solubility, exo-
thermic (heat-generating) chemical reactions, the reflection and refraction of
light, the cohesion of bodies, capillary action—the same phenomena that he
had been unable to explain in strictly mechanical terms. Now they provided
evidence for, and disclosed the power of, the attractions and repulsions he had
added to the properties of matter.

Newton was not entirely finished with aethereal speculations, however.
To the 1717 edition of the *Opticks* he added eight new Queries, in which he
once again proposed the existence of an "aethereal medium" that pervades all
space:

> And is not this Medium the same with that Medium by which Light is refracted
> and reflected, and by whose Vibrations Light communicates heat to bodies . . . ?
> And do not the Vibrations of this Medium in hot Bodies contribute to the intense-
> ness and duration of their Heat? . . . And is not this Medium exceedingly more
> rare and subtile than the Air, and exceedingly more elastick and active? And is it
> not (by its elastick force) expanded through all the Heavens?[11]

This aether possesses some remarkable properties. It is diffused through all
space, being far denser in the empty spaces between the planets than within

10. Ibid., pp. 397–98.
11. Ibid., p. 349.

the planets themselves. As distance from the sun increases, so does the density of this medium.

> And though this Increase of density may at great distances be exceeding slow, yet if the elastick force of this Medium be exceeding great, it may suffice to impel bodies from the denser parts of the Medium towards the rarer, with all that power which we call Gravity. And that the elastick force of this Medium is exceeding great, may be gathered from the swiftness of its Vibrations."[12]

This new aether, unlike the mechanical aether he had proposed earlier, would not resist the motions of the planets: "The exceeding smallness of its Particles may contribute to the greatness of the force by which those Particles may recede from one another, and thereby make that Medium exceedingly more rare and elastick than Air, and by consequence exceeding less able to resist the motions of Projectiles."[13] Newton had not returned to the mechanical aether of the 1670s. Although he used the particles of this later aether to explain almost precisely the same set of phenomena as he had before, they now possessed repulsive forces.

## Explaining Force: Heresy and the Problem of Divine Activity

In addition to natural philosophy, mathematics, and alchemy, Newton devoted a great deal of energy to theological topics. His manuscripts contain at least two and a half million words on theological topics. His heretical theology exacerbated his concerns about introducing and explaining attractive and repulsive forces.

Newton's manuscripts reveal that during the 1670s, he had already broken with the official Trinitarian doctrine of the Anglican Church. Close study of the Bible and the early church fathers convinced him that there is no scriptural evidence for the doctrine of the Trinity. He believed that the Council of Nicaea (325) and especially the church father Athanasius (296–373) had erroneously imposed this doctrine on Christianity in the fourth century. Newton regarded the doctrine of the Trinity as a departure from biblical monotheism. He postulated a God who was both extremely transcendent and immanent. Newton's God was the Lord God of Dominion who created and rules the world by his absolute power. Determined to eliminate the spectre of deism, according to which God created the world and then left it to run on its own according to the laws of nature which he had also created, Newton was therefore deter-

12. Ibid., p. 351.
13. Ibid., p. 352.

mined to establish divine activity in the world, thereby defeating both deism and materialism. Accordingly, his theological papers contain material on the fulfilment of the biblical prophecies—thus exhibiting God's activity in human history—as well as diatribes against the corruption of pristine Christianity by Athanasius, the Council of Nicaea, and the Roman Catholic Church more generally.

Newton believed in the existence of an ancient wisdom that included religion and theology. On this basis, he rejected the doctrine of the Trinity and other "idolatrous" doctrines as corruptions of an originally pure monotheism. His final words in the *Opticks*, the last sentence of Query 31, reassert his belief in the ancient theology (*prisca theologia*): "If the Worship of the false Gods had not blinded the Heathen, then moral Philosophy would have gone farther than to the four Cardinal Virtues; and instead of teaching the Transmigration of Souls and to worship the Sun and the Moon, and dead Heroes, they would have taught us to worship our true Author and Benefactor as their Ancestors did under the Government of *Noah* and his Sons before they corrupted themselves."[14] That corruption was the work, first of the Egyptians, and then of Athanasius and the Catholic Church, which, in Newton's view, had violated monotheism and imposed idolatrous doctrines on an original, pristine religion.

In alchemy, Newton discovered active principles that stimulated his thoughts about the action of forces in the physical world. His recognition that forces and action-at-a-distance presented problems for the mechanical philosophy led to his attempts to find explanations for these forces. Explaining them led him straight back to theology. These twin problems, raised by both the mechanical philosophy and his anti-Trinitarian theology, caused him to contemplate a profound revision of the prevailing mechanical philosophy. Newton's insistence on strict biblical monotheism deeply influenced his natural philosophy—in particular, his conviction that nature is a unity, as the creation of the one God—as well as his theory of universal gravitation.

Newton employed the concept of attractive force as the central concept in the *Principia*. Yet, the mechanical philosophy explicitly ruled out any kind of action-at-a-distance. How did Newton reconcile this apparent contradiction? In the *Principia*, Newton claimed that force is nothing but mathematical shorthand for describing how bodies deviate from inertial motion. In a famous passage in the "General Scholium" appended to the later editions of the *Principia*, he wrote:

14. Ibid., pp. 405–6.

Thus far I have explained the phenomena of the heavens and of our sea by the force of gravity, but I have not yet assigned a cause to gravity. . . . I have not yet been able to deduce from phenomena the reason for these properties of gravity, and I do not feign hypotheses. For whatever is not deduced from the phenomena must be called a hypothesis; and hypotheses, whether metaphysical or physical, or based on occult qualities, or mechanical, have no place in experimental philosophy. . . . And it is enough that gravity really exists and acts according to the laws that we have set forth and is sufficient to explain all the motions of the heavenly bodies and of our sea.[15]

Newton was not being entirely candid in this statement, for he had proposed all sorts of hypotheses—theological, metaphysical, and physical—to explain gravitational attraction and the other attractions and repulsions that he observed throughout the natural world. Nevertheless, by saying that he had not yet been able to discover the cause of gravity or other attractions, Newton implied that the search for such a cause was reasonable. In that case, gravity is not one of the primary qualities of bodies, like extension, impenetrability, and hardness. Thus, he denied that action-at-a-distance is an innate property of bodies, a claim he made explicitly in a letter to the divine, Richard Bentley, who was preparing the first Boyle Lectures:

Tis inconceivable that inanimate brute matter should (without ye mediation of something else wch is not material) operate upon & affect other matter without mutual contact . . . That Gravity should be innate, inherent, and essential to Matter, so that one Body may act upon another at a Distance thro' a Vacuum, without the Mediation of any thing else, by and through which their Action and Force may be conveyed from one to another, is to me so great an Absurdity, that I believe no Man who has in philosophical Matters a competent Faculty of thinking can ever fall into it. Gravity must be caused by an Agent acting constantly according to certain Laws; but whether this Agent be material or immaterial, I have left to the consideration of my readers.[16]

In the *Principia*, Newton had ruled out the possibility that there is a material aether filling space because the resistance to motion caused by such an aether would cause the solar system to wind down very rapidly. The letter to Bentley implies that without the possibility of such a material medium, there must

15. Newton, *Principia*, p. 943.

16. Newton to Richard Bentley, 25 February 1692/93, in *Isaac Newton's Papers and Letters on Natural Philosophy and Related Documents*, ed. I. Bernard Cohen (Cambridge: Harvard University Press, 1958), pp. 302–3.

be an immaterial medium to account for gravitation and for the attractions and repulsions operating between the particles of matter. What might such an immaterial medium be?

In Query 28, Newton proposed that this immaterial medium is God himself, a conclusion reached from an examination of all the order and design evident in the universe.

> Whereas the main Business of Natural Philosophy is to argue from Phaenomena without feigning hypotheses, and to deduce Causes from Effects, till we come to the very first Cause, which certainly is not mechanical; and not only to unfold the Mechanism of the World, but chiefly to resolve these and such like Questions. What is there in places almost empty of Matter, and whence is it that the Sun and Planets gravitate towards one another, without dense Matter between them? Whence is it that Nature does nothing in Vain; and whence arises all that Order and Beauty which we see in the World? . . . How came the Bodies of Animals to be contrived with so much Art, and for what ends were their several Parts? Was the Eye conceived without Skill in Opticks, and the Ear without Knowledge of Sounds? . . . And these things being rightly dispatch'd, does it not appear from Phaenomena that there is a Being incorporeal, living, intelligent, omnipresent, who in infinite Space, as it were in his Sensory, sees the things themselves intimately, and thoroughly perceives them, and comprehends them wholly by their immediate presence to himself? . . . And though every true Step made in this Philosophy brings us not immediately to the Knowledge of the first Cause, yet it brings us nearer to it, and on that account is to be highly valued.[17]

Natural philosophy thus leads directly to knowledge of God. And God's omnipresence performs all the functions that Newton had earlier ascribed to the aether. God is the immaterial cause of the attractions and repulsions that matter exhibits. Where the orthodox mechanical philosophers and the young Newton had attempted to explain the phenomena in terms of invisible mechanical mechanisms, the mature Newton explained them in terms of the direct actions of an infinite, omnipresent, invisible God. Every motion in the universe now became the immediate effect of divine power. Despite his belief in God's literal omnipresence, Newton did not accept pantheism. He insisted on a strict separation between God and the creation.

Not only did Newton's God permeate all space and time, but theology also permeated his philosophy of nature. If such a close alliance between theology and natural philosophy jars modern sensibilities, it seemed perfectly reason-

17. Newton, *Opticks*, pp. 369–70.

able to Newton and his contemporaries. "To treat of God from phenomena," he declared, "is certainly a part of 'natural' philosophy."[18] He opened his first letter to Bentley with the declaration that one of his motivations for writing the *Principia* was to use natural philosophy as a great argument from design. "When I wrote my Treatise about our System, I had an Eye upon such Principles as might work with considering Men, for the Belief in a Deity."[19]

Newton's achievements in physics involved fundamental changes in both the scope and the content of natural philosophy. By mathematizing natural philosophy, he not only created an extraordinarily powerful physical theory, but he also demolished, once and for all, Aristotle's classification of the sciences. This union between mathematics and natural philosophy enabled Newton to fulfill the goals of the mechanical philosophy: to explain the phenomena of the world in terms of matter and motion. The concept of force enabled him to accomplish this goal, but the introduction of attractive and repulsive forces also represented a major departure from one of the major tenets of the mechanical philosophy.

Many of Newton's contemporaries greeted his physics and mathematics enthusiastically, but some found his concept of force troubling. The great German philosopher Gottfried Wilhelm Leibniz (1646–1716)—who had earlier fought with Newton over priority in inventing the calculus—criticized Newton's approach to theology and physics. Leibniz shared a commitment to the centrality of divine providence, but his notion of providence differed sharply from Newton's. Although he adopted a version of the mechanical philosophy, his philosophy also contained several metaphysical principles directly linking it to his theology. He assumed that God created the best of all possible worlds because, as a rational being, God always chooses the best for a good reason. This principle posited the existence of a standard of goodness that exists independently of God's creation, thus committing Leibniz to a form of intellectualism. He also thought that the harmony among things in the world follows from the fact that each individual reflects the entire world at each moment. This harmony implies at least two important consequences: the correspondence between the mind and the body, and miracles that result from the harmony God included in the original creation of the world. Accordingly, a preestablished harmony between mind and body, on the one hand, and between spiritual and physical events on the other, follow from God's initial act of creation. Consequently, God never needs to intervene directly into the

18. Newton, *Principia*, pp. 942–43.
19. Newton to Bentley, 10 December 1692, in Cohen, *Newton's Papers and Letters*, p. 280.

workings of the universe. Divine providence follows from the fact that a perfectly rational God following rational principles—especially the principle of
sufficient reason (the principle that there must be a reason for choosing one
course of action rather than another)—created the best of all possible worlds.

In 1715 and 1716 a controversy about providence erupted between Leibniz
and Newton's spokesman, Samuel Clarke. Leibniz argued that the Newtonian
insistence on divine activity implies that God's workmanship is so imperfect
that he must constantly intervene in nature and repair his work. A better
workman would create a world that would run smoothly forever, without the
need for intervention. Clarke, replying as a Newtonian voluntarist, argued
that Leibniz's account implies an unacceptable limitation on God's freedom
and power because it assumes that God is subject to principles that exist independently of him. In many ways, this debate between the two towering figures
of late-seventeenth-century natural philosophy stands as the culmination of
early modern concerns with the relationship between theology and natural
philosophy.

Despite Newton's groundbreaking work in physics and mathematics, his outlook differed from that of a modern scientist. He believed in an ancient wisdom according to which the works of some of the ancients had prefigured
knowledge of the inverse-square law as well as the truths of monotheistic religion. He thought that the pre-Socratic philosophers, who lived in the sixth and
fifth centuries BC, had known the inverse-square law and that Plato and the
Pythagoreans had foreshadowed his mathematization of nature. He was convinced that Aristotle and, more recently, Descartes had corrupted the ancient
insights.

Newton understood his project to be a double reformation: a reformation
of religion and a reformation of natural philosophy. In both cases he looked
back to ancient traditions with the conviction that their antiquity imparted
legitimacy to his own work. Newton's belief in ancient knowledge—the ancient
geometry, the *prisca theolgia*, and the *prisca sapientia*—belies the common view
that Newton created modern science. Like the Renaissance humanists and
Protestant reformers, Newton looked back to find foundations for his natural philosophy and theology. Modern Newtonianism was the creation of his
eighteenth-century followers, who based their portrait of Newton on a limited
selection of his ideas.

# Epilogue

In 1700 the world—as described by natural philosophers, natural historians, and chymists—looked very different than it had looked in 1500. Particulate matter, inertial motion, and impact replaced Aristotelian matter, form, and the four causes as the ultimate terms of explanation for advocates of the mechanical philosophy. Disciplinary categories shifted: most noteworthy, mathematics became a language for describing physical reality, something that Aristotle had deemed impossible. Explorations of hitherto unknown parts of the world enlarged not only the scope of geographical knowledge but also knowledge of the flora and fauna populating the world. And, for many scholars, the authority of experience, observation, and reason replaced the authority of ancient texts.

Renaissance humanism played a huge role in reviving and renewing the various disciplines, but the several disciplines took separate paths to where they stood in 1700. The revolutionary changes in astronomy resulted from altered methods combined with new observations and yielded an entirely modified view of the cosmos. Changes in the science of motion had more to do with reconceptualization than with observation and experiment. Chymistry, which had always involved a major role for experiment, incorporated ancient theories into new philosophies of nature. The study of living things—both natural history and anatomy and physiology—developed as a revival of ancient methods and tested the veracity of ancient texts. No single description adequately depicts all these developments or the methods that produced them.

These discoveries increased the authority of empirical observation as a reliable source for knowledge of the natural world at the same time that they called for new methods and new accounts of the scope and certainty of human understanding. In different contexts, observation, experiment, and mathematics each played new roles in the search for natural knowledge. Many natural philosophers rejected the Aristotelian goal of *scientia* (certain knowledge of the essences of things) and replaced it with the view that natural philosophy could attain neither certainty nor the knowledge of essences. John Locke, who wrote his *Essay Concerning Human Understanding* (1690) to provide philosophical foundations for the mechanical philosophy, summed up this change of outlook by saying, "Natural Philosophy is not capable of being made a Science."[1] New philosophies of nature and new methods for gaining knowledge replaced Aristotelianism, both its content and method.

Yet, modern science did not follow on the heels of all these developments. Although scholars used the word "science" to describe any number of individual disciplines, they had not yet articulated a general category, science, embracing them all. No one at the time had formulated a general account of the kinds of knowledge and methods that would include natural history, astronomy, medical physiology, chymistry, and the science of motion. Some disciplines described, some explained, some exhibited amazing mathematical sophistication. But no single characterization captured a nature shared by them all.

Science since the nineteenth century and especially in the twentieth and twenty-first centuries has characteristics that had not yet appeared in the early modern period. Natural philosophy is no longer a legitimate disciplinary category. Some of its aspects have fragmented into a host of specialized sciences, such as nuclear physics, biochemistry, cell biology, organic chemistry, computer science, geology, and psychology. Some parts of it no longer rank as sciences at all: theologians now ponder the immortality of the soul, but scientists do not. Similarly, biologists have rejected divine wisdom, power, and goodness as explanations of the order found in the world of living things. And chemists have relegated alchemy to popular culture and the symbolic interpretations of Jungian psychologists and their New Age followers.

The social context of modern science differs profoundly from that of early modern natural philosophers, mathematicians, astronomers, chymists, naturalists, and medical men. With some notable exceptions, such as the Arabic astronomers at Marāgha and Tycho and his team at Hven, practitioners of the sciences in early modern Europe did not work in teams and certainly did not apply for government grants, mentor graduate students and postdoctoral fellows, or publish multi-authored, blindly refereed articles in specialized scientific journals.

These institutional arrangements developed gradually during the eighteenth and nineteenth centuries and then at an enormously accelerated pace in the twentieth and twenty-first centuries. There was no such thing in the

---

1. John Locke, *An Essay Concerning Human Understanding*, ed. Peter H. Nidditch (Oxford: Clarendon Press, 1975), p. 645 (book 4, chap. 12, sect. 10).

early modern period as a professional scientist—someone who is paid for doing research and for replicating himself or herself by training graduate students. All of these social aspects of modern science developed in subsequent centuries.

Where, then, did early modern natural philosophers, astronomers, and medical writers do their work? If they worked in universities—like Galileo at Pisa and Padua, or Newton at Cambridge—they often held chairs in mathematics or one of the sciences, but they did not function in departments or institutes inhabited by like-minded professionals, and they were not in the business of training students who would, in a professional sense, replicate their teachers by carrying on the same traditions in research. If they worked in princely courts, as Kepler did, their contact with other astronomers or natural philosophers consisted largely of correspondence and the publication and reading of books. If, like Copernicus and Gassendi, they earned their livings as ecclesiastical administrators (both men served as canons in their local cathedrals), they pursued their interests in natural philosophy and the sciences on their own time and not as part of their paid positions. A few independently wealthy individuals, like René Descartes and Robert Boyle, did not need employment and pursued their interests on their own. But in most cases, natural philosophers and practitioners of the sciences needed to find either institutional or private patronage to support their work.

Periodicals that specialized in the sciences and natural philosophy came into existence for the first time during the seventeenth century. These publications—such as the *Philosophical Transactions of the Royal Society of London*, the *Acta Eruditorum* in Leipzig, and the *Journal des sçavans* of the Académie des Sciences in Paris—developed from networks of correspondence, often brokered by individuals such as Henry Oldenberg in England or Marin Mersenne (1588–1648) in France.

By the end of the seventeenth century, the landscape of natural knowledge had undergone significant changes. These changes, however, did not yet amount to the emergence of modern science either intellectually or socially. The separation of science and religion, the articulation of a general category called "science," the institutionalization and professionalization of science, and perhaps even a historical phenomenon called "the scientific revolution" or "the rise of modern science" still remained for future generations.

# Suggested Further Reading

## 1 The Western View of the World before 1500

Two general reference works on early modern natural philosophy and the sciences are Wilbur Applebaum, ed., *Encyclopedia of the Scientific Revolution: Copernicus to Newton* (New York: Garland, 2000), and Katharine Park and Lorraine Daston, eds., *The Cambridge History of Science*, vol. 3, *Early Modern Science*, (Cambridge: Cambridge University Press, 2006). For developments in the history of astronomy, see Michael Hoskin, ed., *The Cambridge Concise History of Astronomy* (Cambridge: Cambridge University Press, 1999). Roy Porter presents a general survey of the history of medicine in *The Greatest Benefit to Mankind: A Medical History of Humanity* (New York: Norton, 1997). The complex relations between science and religion in various historical contexts receive careful analysis in John Hedley Brooke, *Science and Religion: Some Historical Perspectives* (Cambridge: Cambridge University Press, 1991).

For a survey of the history of science in antiquity and the Middle Ages, see David C. Lindberg, *The Beginnings of Western Science: The European Scientific Tradition in Philosophical, Religious, and Institutional Context, 600 b.c. to a.d. 1450*, 2nd ed. (Chicago: University of Chicago Press, 2007), and Edward Grant, *The Foundations of Modern Science in the Middle Ages: Their Religious, Institutional, and Intellectual Contexts* (Cambridge: Cambridge University Press, 1996). Roger French and Andrew Cunningham discuss the religious background to medieval natural philosophy in *Before Science: The Invention of the Friars' Natural Philosophy* (Brookfield, VT: Ashgate, 1996). On the translations from Greek into Arabic, see Dimitri Gutas, *Greek Thought, Arabic Culture: The Graeco-Arabic Translation Movement in Baghdad and Early 'Abbāsid Society (2nd–4th / 8th–10th Centuries)* (London: Routledge, 1998). *The Cambridge Companion to Arabic Philosophy*, ed. Peter Adamson and Richard C. Taylor (Cambridge: Cambridge University Press, 2005), provides a good overview of medieval Arabic philosophy.

For a detailed discussion of ancient astronomy, see James Evans, *The History and Practice of Ancient Astronomy* (New York: Oxford University Press, 1998). On the broader cosmological and physical aspects of Ptolemaic astronomy, see Liba Chaia Taub, *Ptolemy's Universe: The Natural Philosophical and Ethical Foundations of Ptolemy's Astronomy* (Chicago: Open Court, 1993). Accounts of astrological theory and practice can be found in Tamsyn Barton, *Ancient Astrology* (London: Routledge, 1994), and S. Jim Tester, *A History of Western Astrology* (Wolfeboro, NH: Boydell Press, 1987). For both the context and a detailed description of Arabic astronomy, see George Saliba, *Islamic Science and the Making of the European Renaissance* (Cambridge: MIT Press, 2007).

Scholarship in English on early alchemy is sparse. On Hellenistic and Byzantine alchemy, see Michèle Mertens, "Graeco-Egyptian Alchemy in Byzantium," in *The Occult Sciences in Byzantium*, ed. Paul Magdaleno and Maria Mauroudi (Geneva: La Pomme d'Or, 2006), pp. 205–30.

Vivian Nutton provides descriptions of the ancient medical traditions in his book *Ancient Medicine* (London: Routledge, 2004). For a through account of ancient natural history, see Roger French, *Ancient Natural History: Histories of Nature* (London: Routledge, 1994).

## 2 Winds of Change

For discussions of the way historians of science have discussed the scientific revolution as well as recent trends in scholarship, see David C. Lindberg and Robert S. Westman, eds., *Reappraisals of the Scientific Revolution* (Cambridge: Cambridge University Press, 1990), and Margaret J. Osler, ed., *Rethinking the Scientific Revolution* (Cambridge: Cambridge University Press, 2000).

Wide-ranging discussions of Renaissance humanism can be found in Anthony Grafton, with April Shelford and Nancy Siraisi, *New Worlds, Ancient Texts: The Power of Tradition and the Shock of Discovery* (Cambridge: Belknap Press of Harvard University Press, 1992), and Jill Kraye, ed., *The Cambridge Companion to Renaissance Humanism* (Cambridge: Cambridge University Press, 1999). On philosophy in the Renaissance, see Brian P. Copenhaver and Charles B. Schmitt, *Renaissance Philosophy* (Oxford: Oxford University Press, 1992); Richard Popkin, *The History of Scepticism from Savonarola to Bayle*, rev. and expanded ed. (Oxford: Oxford University Press, 2003); and Charles B. Schmitt, Quentin Skinner, and Eckhard Kessler, eds., *The Cambridge History of Renaissance Philosophy* (Cambridge: Cambridge University Press, 1988). An account of the development and impact of printing is provided by Elizabeth L. Eisenstein, *The Printing Press as an Agent of Change* (Cambridge: Cambridge University Press, 1979). On developments in anatomy, see Andrew Cunningham, *The Anatomical Renaissance: The Resurrection of the Anatomical Projects of the Ancients* (Brookfield, VT: Ashgate, 1997). See Frances A. Yates, *Giordano Bruno and the Hermetic Tradition* (Chicago: University of Chicago Press, 1964), for a discussion of the Hermetic tradition.

On the conditions of life during the Reformation, see Andrew Cunningham and Ole Peter Grell, *The Four Horsemen of the Apocalypse: Religion, War, Famine and Death in Reformation Europe* (Cambridge: Cambridge University Press, 2000). Two books that provide the details of the history of the Reformation and its consequences are Diarmaid MacCulloch, *The Reformation: A History* (New York: Viking, 2003), and Steven Ozment, *The Age of Reform: 1250–1550* (New Haven: Yale University Press, 1980).

For an account of how Europeans reacted to the flora and fauna of the New World, see Miguel de Asúa and Roger French, *A New World of Animals: Early Modern Europeans on the Creatures of Iberian America* (Burlington, VT: Ashgate, 2005).

For lucid explanations of the technical aspects of the Copernican revolution, see Thomas S. Kuhn's *The Copernican Revolution: Planetary Astronomy in the History of Western Thought* (Cambridge: Harvard University Press, 1957). In *The Sleepwalkers: A History of Man's Changing Vision of the Universe* (London: Penguin, 1964; first published 1959), Arthur Koestler gives a discursive account of the contributions of Copernicus and Kepler; and Peter Barker provides the historical background to Copernicus in his article "Copernicus and the Critics of Ptolemy," *Journal of the History of Astronomy* 30 (1999): 343–58. The relationship between Kepler's theology and astronomy is discussed in Peter Barker and Bernard R. Goldstein, "Theological Foundations of Kepler's Astronomy," *Osiris* 16 (2001): 88–113. On the religious implications of and reactions to Copernican astronomy, see Kenneth J. Howell *God's Two Books: Copernican Cosmology and Biblical Interpretation in Early Modern Science* (Notre Dame, IN: University of Notre Dame Press, 2002).

## 3 Observing the Heavens

For a brief account of Galileo's life and works, along with translations of some of his most important writings, see Maurice A. Finocchiaro, ed., *The Essential Galileo* (Indianapolis: Hackett, 2008). Dava Sobel presents an engaging account of Galileo's life and works in *Galileo's Daughter: A Historical Memoir of Science, Faith, and Love* (Toronto: Viking, 1999).

A vast literature exists on Galileo and the Church. The following books touch on key issues surrounding his ideas, his trial, and his condemnation by the Inquisition: Richard J. Blackwell, *Galileo, Bellarmine, and the Bible, Includ-*

*ing a Translation of Foscarini's Letter on the Motion of the Earth* (Notre Dame, IN: University of Notre Dame Press, 1991); Annibale Fantoli, *Galileo: For Copernicanism and for the Church*, trans. George V. Coyne, 2nd ed. (Rome: Vatican Observatory Publications, 1996); and Ernan McMullin, ed., *The Church and Galileo* (Notre Dame, IN: University of Notre Dame Press, 2005).

On astrology, see Patrick Curry, ed., *Astrology, Science and Society* (Wolfeboro, NH: Boydell Press, 1987) and Anthony Grafton, *Cardano's Cosmos: The World and Works of a Renaissance Astrologer* (Cambridge: Harvard University Press, 1999).

## 4  Creating A New Philosophy of Nature

On the Aristotelian background to the mechanical philosophy, see Dennis Des Chene, *Physiologia: Natural Philosophy in Late Aristotelian and Cartesian Thought* (Ithaca: Cornell University Press, 1996). Descartes' approach to the mechanical philosophy is summarized by Stephen Gaukroger, *Descartes' System of Natural Philosophy* (Cambridge: Cambridge University Press, 2002). For the history of debates about the existence of the void, see Edward Grant, *Much Ado about Nothing: Theories of Space and the Vacuum from the Middle Ages to the Scientific Revolution* (Cambridge: Cambridge University Press, 1981). On the controversy between Robert Boyle and Thomas Hobbes about whether Boyle's experiments established the existence of the void, and more generally about the utility of an experimental approach, see Steven Shapin and Simon Schaffer, *Leviathan and the Air-Pump: Hobbes, Boyle, and the Experimental Life* (Princeton: Princeton University Press, 1985).

Brian P. Copenhaver provides a history of a common list of occult qualities in his "A Tale of Two Fishes: Magical Objects in Natural History from Antiquity through the Scientific Revolution," *Journal of the History of Ideas* 52 (1991): 373–98. The role of occult qualities in the mechanical philosophy is discussed by Keith Hutchison, "What Happened to Occult Qualities in the Scientific Revolution?" *Isis* 73 (1982): 233–53, and John Henry, "Occult Qualities and the Experimental Philosophy: Active Principles in Pre-Newtonian Matter Theory," *History of Science* 24 (1986): 335–81.

For the latest views on the relationship between science and religion in early modern Europe, see Ronald L. Numbers, *Galileo Goes to Jail and Other Myths about Science and Religion* (Cambridge: Harvard University Press, 2009). Religion and theology figured significantly in early modern science and philosophy, as discussed in the following: on the role of biblical interpretation and its relationship to the development of the sciences, Peter Harrison, *The Bible, Protestantism, and the Rise of Natural Science* (Cambridge: Cambridge University Press, 1998); on the role of theological presuppositions regarding divine activity and providence in the mechanical philosophy, Margaret J. Osler, *Divine Will and the Mechanical Philosophy: Gassendi and Descartes on Contingency and Necessity in the Created World* (Cambridge: Cambridge University Press, 1994); on divine purpose in natural philosophy, Osler, "Whose Ends? Teleology in Early Modern Natural Philosophy," *Osiris* 16 (2001): 151–68. For the interconnections between theology and natural philosophy in Robert Boyle's thought, see Jan W. Wojcik, *Robert Boyle and the Limits of Reason* (Cambridge: Cambridge University Press, 1997). See also Peter Harrison, *The Fall of Man and the Foundations of Science* (Cambridge: Cambridge University Press, 2007).

## 5 Shifting Boundaries

On the disciplinary status of the science of mechanics, see Alan Gabbey, "Between *Ars* and *Philosophia Naturalis*: Reflections on the Historiography of Early Modern Mechanics," in *Renaissance and Revolution: Humanists, Scholars, Craftsmen, and Natural Philosophers in Early Modern Europe*, ed. J. V. Field and A. J. L. Frank James (Cambridge: Cambridge University Press, 1993). The development of mechanics is described in Domenico Bertoloni Meli, *Thinking with Objects: The Transformation of Mechanics in the Seventeenth Century* (Baltimore: Johns Hopkins University Press, 2006). On the metaphysical implications of the new science of motion, see Margaret J. Osler, "Galileo, Motion, and Essences," *Isis* 64 (1973): 504–9.

For background on medieval optics, see David C. Lindberg, *Theories of Vision from Al-Kindi to Kepler* (Chicago: University of Chicago Press, 1976). The seventeenth-century developments in optics are discussed in A. I. Sabra, *Theories of Light (from Descartes to Newton)* (London: Oldbourne, 1967). On Newton's experiment on colors, see Richard S. Westfall, "The Development of Newton's Theory of Colors," *Isis* 53 (1962): 339–58.

## 6 Exploring the Properties of Matter

A general history of chemistry is given in Trevor H. Levere's book *Transforming Matter: A History of Chemistry from Alchemy to the Buckyball* (Baltimore: Johns Hopkins University Press, 2001). For an authoritative account of the history of alchemy, see Lawrence M. Principe, *The Secrets of Alchemy* (Chicago: University of Chicago Press, 2011); also see Bruce T. Moran, *Distilling Knowledge: Alchemy, Chemistry, and the Scientific Revolution* (Cambridge: Harvard University Press, 2005). The philosophical implications of alchemy are explored by William R. Newman, *Promethean Ambitions: Alchemy and the Quest for Perfection in Nature* (Chicago: University of Chicago Press, 2004).

On Paracelsus and his followers, see Allen G. Debus, *The Chemical Philosophy: Paracelsian Science and Medicine in the Sixteenth and Seventeenth Centuries*, 2 vols. (Chicago: University of Chicago Press, 1977). William R. Newman discusses the alchemical background to seventeenth-century corpuscularian alchemy in *Atoms and Alchemy: Chymistry and the Experimental Origins of the Scientific Revolution* (Chicago: University of Chicago Press, 2006). On Robert Boyle's devotion to alchemy and his intellectual development, see William R. Newman and Lawrence Principe, *Alchemy Tried in the Fire: Starkey, Boyle, and the Fate of Helmontian Chymistry* (Chicago: University of Chicago Press, 2002), and Lawrence M. Principe, *The Aspiring Adept: Robert Boyle and His Alchemical Quest* (Princeton: Princeton University Press, 1998).

## 7 Studying Life

For general background on natural history, see Brian W. Ogilvie, *The Science of Describing: Natural History in Renaissance Europe* (Chicago: University of Chicago Press, 2006), and Nicholas Jardine, James A. Secord, and E. C. Spary, *The Cultures of Natural History* (Cambridge: Cambridge University Press, 1995). On collectors, museums, and patronage, see Paula Findlen, *Possessing Nature: Museums, Collecting, and Scientific Culture in Early Modern Italy* (Berkeley and Los Angeles: University of California Press, 1994). On the Accademia dei Lincei, see David Freedberg, *The Eye of the Lynx: Galileo, His Friends, and the Beginnings of Modern Natural History* (Chicago: University of Chicago Press, 2002). Early views about fossils are discussed in Martin J. S. Rudwick, *The Meaning of Fossils: Episodes in the History of Paleontology*, 2nd ed. (New York: Science History Publications, 1976).

Roger French discusses Harvey's Aristotelianism in *William Harvey's Natural Philosophy* (Cambridge: Cambridge University Press, 1994). On Harvey's followers in England and their experimental practice, see Robert G. Frank, Jr., *Harvey and the Oxford Physiologists* (Berkeley and Los Angeles: University of California Press, 1980).

On the Aristotelians' and Descartes' views of the soul, see two books by Dennis Des Chene, *Life's Form: Late Aristotelian Conceptions of the Soul* (2000) and *Spirits and Clocks: Machine and Organism in Descartes* (2001), both published by Cornell University Press, Ithaca, NY. For Gassendi's arguments for the immortality of the soul, see Margaret J. Osler, "Baptizing Epicurean Atomism: Pierre Gassendi on the Immortality of the Soul," in *Religion, Science, and Worldview: Essays in Honor of Richard S. Westfall*, ed. Margaret J. Osler and Paul Lawrence Farber (Cambridge: Cambridge University Press, 1985), pp. 163–84.

## 8 Rethinking the Universe

A general account of Newton's natural philosophy is provided in Betty Jo Teeter Dobbs and Margaret C. Jacob, *Newton and the Culture of Newtonianism* (Atlantic Highlands, NJ: Humanities Press, 1995). The definitive biography of Newton is Richard S. Westfall, *Never at Rest: A Biography of Isaac Newton* (Cambridge: Cambridge University Press, 1980). J. E McGuire and Martin Tamny provide a modern transcription of Newton's undergraduate notebook in *Certain Philosophical Questions: Newton's Trinity Notebook* (Cambridge: Cambridge University Press, 1983). For the reasoning in the *Principia*, see Dana Densmore, *Newton's Principia: The Central Argument*, 3rd ed. (Santa Fe, NM: Green Lion, 2003).

Newton's speculative writings on the nature of matter, the concept of force, and other topics are available in a volume edited by I. Bernard Cohen, *Isaac Newton's Letters and Papers on Natural Philosophy* (Cambridge: Harvard University Press, 1958). For a detailed account of the development of Newton's concept of force as well as the development of the concept during the seventeenth century, see Richard S. Westfall, *Force in Newton's Physics: The Science of Dynamics in the Seventeenth Century* (New York: Elsevier, 1971).

On Newton's alchemy and its relationship to his heterodox religion, see B. J. T. Dobbs, *The Janus Faces of Genius: The Role of Alchemy in Newton's Thought* (Cambridge: Cambridge University Press, 1991). Newton's heresy and its relationship to his physics are discussed in two articles by Stephen D. Snobelen: "'God of Gods and Lord of Lords': The Theology of Isaac Newton's General Scholium to the *Principia*," *Osiris*, 2nd ser., 16 (2001): 169–208, and "'The True Frame of Nature': Isaac Newton, Heresy, and the Reformation of Natural Philosophy," pp. 223–62 in *Heterodoxy in Early Modern Science and Religion*, ed. John Brooke and Ian Maclean (Cambridge: Cambridge University Press, 2005).

# Index

Abelard, Peter, 4
Accademia dei Lincei, 65
Acosta, José, on natural history of the New
  World, 135
*Acta Eruditorum*, 167
action-at-a-distance: and force, 154; Gassendi
  on, 81; the mechanical philosophy on, 86;
  in Newton, 155–7
active principles, Newton on, 160
aether: and the mechanical philosophy, 155; in
  Newton, 154, 157–8
afterlife, Boyle on, 92
air: Boyle on, 129; as element in Aristotle, 9
air-pump, Boyle on, 129, 143
Albertus Magnus, on faith and reason, 12
alchemy: in Alexandria, 21; Arabic, 21; Roger
  Bacon on, 23; Boyle on, 92, 126–7; and
  chemistry, 118; Geber on, 23; Locke on,
  127; and matter, 118; Newton on, 127, 157,
  160; origins of, 21; John of Rupecissa on,
  23; Van Helmont on, 126
Alciati, Andrea, on emblems, 133–4
Alexandria: alchemy in, 21; anatomy in, 27, 33
Alhazen: on astronomy, 18–9; on optics, 103,
  110–1
al-Khwārismī, 40
al-Kindī, 11
al-Tūsī, 19
ammonites: Lister on, 139; Voltaire on, 139
anastomoses: Galen on, 28; Harvey on, 142
anatomy, 132, 165; in Alexandria, 27, 33;
  Fabricius on, 141; Galen on, 34, 35; in the
  Middle Ages, 33; nomenclature of, 36; in
  the Renaissance, 33; Vesalius on, 34–6
angels: Boyle on, 92; Gassendi on, 80
animism, in Kircher, 137
Aquinas, Thomas: and Aristotle, 5, 12, 39; on
  divine will, 90; on the eucharist, 12; on
  faith and reason, 12

Archimedes, 33; Galileo and, 96; and law of
  the lever, 96; method of, 96
Aristarchus of Samos, 49
Aristotelianism, 165; Galileo's challenges to,
  62–3; mechanical philosophy's rejection of,
  77–78, 84–85, 99–100, 117, 166; on vision,
  103
Aristotle, 3, 6–11; on cause, 6, 7; on classification of disciplines, 2, 6; on the common
  sense, 109; *De anima*, 145; on elements, 9,
  20; natural books by, 6, 10, 145; on natural
  history, 23–24, 132, 136; Paracelsus on, 119;
  rejection of, by chemical philosophers, 121;
  on soul, 9–10, 145; on stones and metals,
  134; on teleology, 7–8, 23–24
Arouet, François Marie. *See* Voltaire
astrology, 72–5; Augustine on, 72; Babylonian, 13; Bellanti on, 73; Calvin on, 74;
  judicial, 72; Kepler on, 55, 75; Melancthon
  on, 74; natural, 72; in Paracelsus' medicine, 120; Pico on, 72–3; Pontano on, 73;
  Ptolemy on, 17; Tycho on, 75
astronomy: Arabic, 18, 43; Averroes on, 42–3;
  Babylonian, 13; Copernicus on, 43–51;
  geo-heliocentric, 53–4; Greek, 13–15; heliocentric, 43–60; Hipparchus on, 15; and
  mathematics, 14, 20; methods of, 165;
  observations in, 52–3, 165; Peurbach on,
  42–3, and physics, 50–1, 53, 55, 57, 59–60,
  69; Plato on, 13–14; Ptolemy on, 3, 15–17,
  18–19, 42–3; Regiomontanus on, 42–3
Athanasius, Newton on, 160
atheism, 89
atomism: Epicurus on, 77; Gassendi on, 78,
  80–1; Lucretius on, 32
atomists: on matter, 20, 78–81; on vision,
  103
atoms: Epicurus on, 77; Gassendi on, 78,
  80–1; properties of, 81

Augustine, Saint: on astrology, 72; on biblical
interpretation, 39, 66–7
Averroes, 11–2; criticism of Ptolemaic astronomy by, 42–3
Avicenna: on medicine, 28–9; on mercury-
sulfur theory, 22; Paracelsus on, 119; on
transmutation, 22

Bacon, Francis, 136–7
Bacon, Roger, on alchemy, 23
Barbaro, Ermolao, and natural history, 132–3
Barberini, Maffeo (Pope Urban VIII), 69
barometer: Boyle on, 129; and chymistry,
130; Descartes on, 80; Gassendi on, 79–80;
in mechanical philosophy; Pascal on, 79;
Roberval on, 80; Torricelli on, 79
Bauhin, Gaspard, on natural history, 134
beavers: *Physiologus* on, 26; Pliny on, 24–6;
Topsell on, 138
Becher, Johann Joachim, on phlogiston, 130
Bellanti, Lucio, on astrology, 73
Bellarmine, Robert, 64; on Copernican
astronomy, 68–9; letter by, to Foscarini, 69
Bentley, Richard, 93, 161
Bessel, Friedrich Wilhelm, and parallax, 50
Bible, 1; authority of, 37; and chemical philosophy, 121; on earth's age, 137; Galileo on,
66–7; on interpretation of, 38–40, 66–7;
and science, 38–40; on species, 138–9;
translation of, 37–8
biology, as defined discipline, 132, 146
Bīrūnī, 40
blood: circulation of, 142–3; color of, 144; in
humor theory, 26
Boyle, Robert: on air, 129, 167; on air-pump,
129; and alchemy, 92, 126–7; on angels,
92; on color, 111; and Descartes, influence
of, 128; on elements, 127–8; and Gassendi,
influence of, 128; on God, 92, 145; on limits of human reason, 92; on matter, 127–9;
and the mechanical philosophy, 88, 128–9;
More on, 91; and natural philosophy, 126;
Newton on, 154; on *Origin of Forms and
Qualities*, 128; and philosophers' stone,
126; on qualities, 129; and respiration
experiments, 143; *Sceptical Chymist*, 127;
on the soul, 145; and Starkey, 126, 128; on
theology, 92, 145; and Van Helmont, influence of, 126; on void, 129; on willow tree
experiment, 128

Boyle Lectures, 93, 161
Buridan, Jean, 12

Cabala, 32
cabinets of curiosities, 136
Caccini, Tommaso, attack on Galileo by, 66,
68
calcination, Stahl on, 130
calculus, Newton on, 149, 151
calendar reform, 43, 114n14
Calvin, Jean, 37; on astrology, 74
cardiovascular system, 144; Colombo on,
140–1; Galen on, 28, 140–1; Harvey on,
141, 143; Ibn al-Nafis on, 140
Casaubon, Isaac, 31–32
Castelli, Benedetto, 66
Catholic Church: and natural philosophy,
167; Newton on, 160
cause: Aristotle on, 6–7; Gassendi on, 81;
God as first, 81
censorship, ecclesiastical, 68–9; and Descartes, 84; and Galileo, 70–1; and Kepler,
69
Cesi, Frederico, 65
chemical philosophy, 118–24
chemistry, 126–30; and alchemy, 118; origins
of, 21
chymistry: definition of, 118; experiments in,
165; and the mechanical philosophy, 130
Cicero, Marcus Tullius, 31
circular motion: Descartes on, 102; Huygens
on, 101–2; Newton on, 148
Clarke, Samuel, 93; against Leibniz, 164
classification of plants and animals: Linnaeus
on, 140; in Newton, 153; Ray on, 140
Collegio Romano, 64
Colombo, Realdo, on pulmonary circulation,
140–1
color: Aristotelians on, 114; Descartes on, 105,
114; mechanical philosophers on, 111; Newton on, 113–7, 149
Columbus, Christopher, 40–1
combustion, Stahl on, 130
comets: Flamsteed on, 75; Halley on, 75;
Newton on, 75–6, 153; as portents, 75;
Tycho on, 53
common sense, the, 109
Condemnation of 1277, 12
Conduitt, John, 147
Copernican astronomy, reception of, 52, 55

Copernicus, Nicholas, 43–51, 167; *De revolutionibus*, 44
cosmology, 52–3, 60; Aristotle on, 8; Copernicus on, 45–6; Descartes on, 83; Hermetic, 32; medieval discussions of, 12; Ptolemy on, 17; rejection of Aristotelian, 99
Council of Nicaea, Newton on, 160
Council of Trent, 38, 52, 67
Counter-Reformation, 38
creation: Christian view of, 5; Paracelsus' chemical explanation of, 120
Cremonini, Cesare, 63

deism, Newton on, 158
demons, Gassendi on, 80
*De revolutionibus*, 44; banning of, by the Church, 68
Derham, William, 93
Descartes, René, 78, 81–4, 87–8, 91, 167; on animals, 145; on the barometer, 80; on circular motion, 102; on colors, 105, 111; on the common sense, 109; experimental model of rainbow by, 111; force in, 103; on impact, 100–1; on inertia, 100; influence of Kepler on, 104; influence of, on Boyle, 128; on kinematics, 103; and laws of motion, 82–3, 100, 104; on light, 83, 104–9, 111–2; on the magnet, 87–8; on motion, 95; Newton on, 154; on optics, 104–8; on the pineal gland, 109; on the rainbow, 109–12; on reflection, 106–8; on refraction, 106, 108, 112; on the soul, 145; on the telescope, 109; on vortices, 112
design, argument from, 161
Dioscorides, 24; natural history by, 132–3
disciplinary boundaries, 3, 14, 50, 72, 95, 132, 145–6, 165–6; Aristotle on, 2, 162; and chymistry, 131; and mixed mathematics, 15, 29, 52, 94, 117; Newton on, 162; in university curricula, 13
disease: in Greek medicine, 120; Paracelsus on, 120
dissection, 27, 33–4
divine activity, Newton on, 160
divine power, 91, 164
divine will, 91, 164

earth: at center of the cosmos, 16; history of, 2, 137

earth's motions: 45–6, 48, 69; Galileo on, 95, 98; objections to, 49–50, 98–9
eccentric, 42; as equivalent to epicycle, 15
Eco, Umberto, on weapon salve, 122n3
educational reform, and chemical philosophy, 121
Eirenaeus Philalethes, 126
elements: Aristotle on, 9; Boyle on, 127–8; Descartes on, 104; Paracelsus on, 120; Van Helmont on, 125
emblems, Alciati on, 133–4
Epicurus, 77–8; on void, 79
epicycle, 42; as equivalent to eccentric, 15
equant, 42, 52; Alhazen's rejection of, 19; and Copernicus, 48; in Ptolemy, 18–19
eucharist, 12; Descartes on, 84
Euclid, 33, 96; on vision, 103
experiments: alchemical, 118; and Boyle, 128–9, 143; by Harvey, 143; Lower's, on respiration, 143–4; in Newton on colors, 115, 154, 156; in Paracelsus, 120; quantitative approach to, 125; and rainbow model, 111; Theodoric on, 111; in Van Helmont, 124–5, 128
exploration, voyages of, 40–1, 165; motives for, 135; and natural history, impact on, 132

Fabricius of Aquapendente, on valves in the veins, 141
faith and reason, in the Middle Ages, 12
Fermat, Pierre de, on law of refraction, 113
Ficino, Marsilio, 31
Flamsteed, John, on comets, 75
Fludd, Robert: as chemical philosopher, 121; on weapon salve, 121–3
fluxions, Newton's method of, 149, 151
force, 95; Huygens on, 101–3; Kepler on, 59; and the mechanical philosophy, 102, 147, 149, 154; Newton on, 147–9, 156, 160
Foscarini, Paolo Antonio, 69
fossils: Gessner on, 134; Hooke on, 138; location of, 138–9; Kircher on, 137; Steno on, 137–8; Voltaire on, 139
Foster, William, attack by, on Fludd and weapon salve, 122–3
freedom, divine, 90–1, 164; Boyle on, 92
Froben, Johannes, 118–9

Galen, 3, 26–8, 33; on anatomy, 34, 35; on cardiovascular system, 28, 141, 144; down-

Galen *(cont.)*
fall of, 144; method of, 143; Paracelsus on,
119; physiology of, 141; rejection of, by
chemical philosophers, 121; and teleology,
27, 36, 144
Galileo Galilei: and Archimedes, 96; and the
Church, 65–72; criticism of Aristotelian
cosmology by, 62; *De motu,* 96, 98; *Dialogue on the Two Chief World Systems,* 70–1,
97–8; *Discourses on Two New Sciences,*
71–2, 97; on earth's motions, 70, 95–9;
on inertia, 98–9; and Kepler, 55; 65; on
kinematics, 98, 102; on law of falling bodies, 97; *Letter to the Grand Duchess,* 66–7;
on motion, 61, 96–9; and patronage, 63;
on science and scripture, 66–72; *Siderius
nuncius,* 62–3; and the telescope, 61–4; on
the tides, 70; trial of, 71; on uniformity of
nature, 65, 70; on void, 97
gardens, medical, 133–4
Gassendi, Pierre, 167; atomism of, 78, 80–1;
Boyle, influence on, 128; and Hobbes,
89–90; on inertia, 99; on knowledge,
91; on light, 85–6; Newton on, 154; on
qualities, 84–5; on theology, 78, 80, 84–5,
89–90, 145; on void, 78; on the weapon
salve, 123–4
Geber, on alchemy, 22–3
geography, 165; Arabic writers on, 49;
Ptolemy on, 40–1
geo-heliocentric astronomy, 53
Gessner, Conrad, 133
Gilbert, William, 87
Glanvill, Joseph, 91
God, 1; argument for, from design, 161;
attributes of, 90; Boyle on, 92, 145; Clarke
on, 164; Descartes on, 83, 91; freedom
of, 91, 164; Gassendi on, 78, 80, 91; and
gravity, 161; Hernández on, 135; Kepler
on, 57–9; knowledge of, 37; Leibniz on,
162, 164; and the mechanical philosophy,
89; Newton on, 158, 160–2, 164; power of,
91, 164; Ray on, 140; as source of motion,
89; Topsell on, 137; Van Helmont on,
124–5
gold-making, 21, 127
gravity, 95; Descartes on, 83; inverse-square
law of, 148; Newton on, 148–51, 154, 161;
Grimaldi, Francesco, on color, 111
Guenther von Andernach, 34

Halley, Edmond: on comets, 75; and
Newton, 150
Harriot, Thomas, and the telescope, 61
Harvey, William: on anatostomoses, 142; on
cardiovascular system, 142–3
heliocentric astronomy: Aristarchus on, 49;
Copernicus on, 43–51
herbaria, 134
Hermes Trismegistus, 31
Hermetic writings, 31–2; and chemical
philosophy, 121
Hernández, Francisco: on natural history of
New Spain, 135
Hero of Alexandria, 79
Hipparchus on equivalence of eccentric and
epicycle, 15–6
Hippocrates, 3; theory of humors of,
26–7
Hobbes, Thomas: determinism of, 89; and
Gassendi, 89–90; materialism of, 84, 89;
Newton on, 154
Hooke, Robert: on color, 111; and correspon-
dence with Newton, 149; on fossils, 138;
and inverse-square law, 149; and Newton's
theory of colors, criticism of, 117
Horky, Martin, 64
*horror vacui,* 79–80, 129
humors, theory of, 26–7, 33, 120
Huygens, Christiaan: on light, 113–5; on
motion, 101–2

Ibn al-Nafis, on pulmonary circulation,
140
illustration: medical, 34–6; in natural history,
133–5
impact, 94; Descartes on, 100–1; Gassendi
on, 81; Hugyens on, 101
indigenous knowledge, Hernández' use of, 135
inertia, 94, 117, 165; Descartes on, 100; Galileo on, 98–9; Gassendi on, 99; Newton
on, 148
Inquisition, the, and Galileo, 68, 71
Islam, rise of, 4

Jabīr, and alchemy, 21–2
Jesuits, 38, 117; on astronomy, 64; and Des-
cartes, 83; and Galileo, 71; and Kircher,
137; and Tycho's astronomy, 69
*Journal des sçavans,* 167
Jupiter, satellites of, 62–3

Kepler, Johannes, 167; and Alhazen's optics, influence of, 104; on astrology, 55, 75; on astronomy, 54–60; books by, banning of, 69; on Galileo's telescopic discoveries, 63; on God, 54, 57–8; influence of, on Descartes, 108–9; on mathematics and physics, 104; on optics, 104; and Osiander's preface, 44; and the telescope, 104
Kepler's Laws, 57–9; and inverse-square law, 151; and Newton, 148
kinematics: definition of, 97; in Descartes, 101, 103; Galileo on, 97, 102; in Huygens, 101, 103
Kircher, Anasthasius, 137

lapidifying virtue, Kircher on, 137
Lavoisier, Antoine-Laurent, on chemistry, 130
law of falling bodies: Galileo on, 97; and inverse-square law, 152
law of reflection of light, Descartes on, 106
law of refraction of light: Descartes on, 106, 108–9, 112; Fermat on, 113; Huygens on, 113, 115
law of the lever, Archimedes on, 96
law of universal gravitation, 153
laws: Boyle on, 92; and God, 90; Kepler's, 57–9; and Newton, 148, 151, 153
laws of motion, Descartes', 82–3, 100, 104; Newton on, 150–1
Leibniz, Gottfried Wilhelm, 162; and correspondence with Clarke, 164; on God, 164
lenses, Descartes on, 109
Leoniceno, Niccolò, on Pliny, 132–3
light, 83; Alhazen on, 103; composition of, 115; Descartes on, 83, 104–9, 111–2; Gassendi on, 85–6; Greeks on, 96; Huygens on, 113–5; Kepler on, 104, 109; Newton on, 113–7, 149; Pecham on, 103; refraction of, 112–3, 115; Witelo on, 103
Linnaeus, Carolus, 140
Linus, Francis, and criticism of Newton's theory of colors, 117
Lipperhey, Hans, and the telescope, 61
Lister, Martin, on fossils, 139
Locke, John: on alchemy; 127; on natural philosophy and science, 165–6
Lorini, Niccolò, and attack on Galileo, 65–6
Lower, Richard, and experiments on respiration, 143–4
Loyola, Ignatius, 38

Lucretius, 32
lunar theory, Newton on, 153
Luther, Martin, 37

macrocosm and microcosm, 32; and the chemical philosophy, 121; in Fludd, 122; in Paracelsus, 119–20
Maestlin, Michael, 54–5
Magellan, Ferdinand, 41
Magini, Giovanni Antonio, 64
magnets: Descartes on, 87–8; Gilbert on, 88
Malpighi, Marcello, on capillaries, 143
Marāgha observatory, 19, 166
Maria the Jewess, 21
mass, 95
materialism: in Hobbes, 89; and the mechanical philosophy, 132, 144
mathematics: Aristotle on, 6; and astronomy, 19; in Huygens, 101; mixed, 15, 29, 52, 94, 117; Newton on, 149, 151; and physics, 100–1; practical, 41; Pythagorean, 55
matter: in alchemy, 118; Aristotle on, 6–10, 20; atomists on, 20; Becher on, 130; Boyle on, 127, 129; Descartes on, 82, 84, 87, 104; Gassendi on, 80–1, 84–5; Huygens on, 101; Jabīr on, 21–2; Paracelsus on, 120; Rhazes on, 22; Stahl on, 130
mechanical models, in Descartes, 105–8
mechanical philosophy, the; Boyle on, 88, 128–9; Descartes on, 82–4, 87–8; and force, 102, 147; Gassendi on, 78–82, 84–6, 89–91; Hobbes on, 84; and light, explanation of, 105–7; limits of, 145; and materialism, danger of, 132, 144; and Newton, 99, 154–5, 160, 162; on qualities, 84–5; and theology, 89
mechanics, definition of, 94–5
medicine: Avicenna on, 28; and chemical philosophy, 121; Greeks on, 3, 26–8
Melancthon, Philip, 52; on astrology, 74
mercury: and metals, theory of, 22; Paracelsus' medicinal use of, 120; in transmutation, use of, 127
Mersenne, Marin, 167
metals, mercury sulfur theory of, 22
method, 165; Archimedes on, 96; Aristotle on, 10; Galen on, 143; Harvey on, 143; inductive, in Bacon, 136; Jabīr on, 22
microscope, 81; Malpighi's use of, 143
miracles, 12; Boyle on, 145; of Joshua, 66; Leibniz on, 162

mixed mathematics, 15, 29, 52, 94, 117
moon: Galileo's observations of, 62, Newton
on theory of, 153
More, Henry, 91
motion, 50, 82, 165; Aristotle on, 9; cause of,
in mechanical philosophy, 95–6; circular,
94, 101–2; Descartes on, 82, 95; Galileo on,
95; natural and violent, 96–8; projectile,
98; relativity of, 45, 50, 101; science of,
94–103
museums, 136

natural history, 132–6, 165; ancient, 23–6; and
explorations, impact of, 41–2, 132
natural motion: Aristotle on, 8–9; Galileo
on, 96–7
natural philosophy, 161, 165; Aristotelian, 3, 5,
136; definition of, 2, 94; Locke on, 165–6;
Newton on, 162; and theology, 29
natural theology, Ray on, 140
nature: in Aristotelianism, 89; in early modern period, 137–40; in the Renaissance,
132–6
Neoplatonism: in Fludd, 121; on fossils, 134
Newton, Isaac, 167; on action-at-a-distance,
155–7, 160; on active principles, 156–7, 160;
on aether, 154–5; on alchemy, 126–7, 147,
149, 157, 160; "annus mirabilis" of, 149; on
argument from design, 162; and Aristotelianism, rejection of, 154; on Athanasius,
160; on biblical prophecies, 160; and Boyle,
154; and calculus, 149, 151; on colors, 114–7,
149; on comets, 75–6, 153; on Council of
Nicaea, 160; and experiment on colors,
115–6; and experiments, 154, 156; on fluxions, 149, 151; on force, 147–9, 150–1, 154,
156, 160; on God, 158, 160, 162; on gravity, 148–51, 154, 161; and Halley, 150; and
Hooke, 117, 149–50; on impact, 148; and
inertia, 148, 156–7; and inverse-square law,
149–50, 152; and Kepler's Laws, 148, 150–1;
and Keynes, 148; and law of falling bodies,
152; laws of motion of, 150–1; on light, 113–
7, 149; on lunar theory, 153; manuscripts of,
147–8; on mathematics, 149, 151; and the
mechanical philosophy, 88, 99, 148, 154–6,
160, 162; method of, 147, 154–5, 157, 160,
162; on natural philosophy, 162; Opticks,
147; on orbiting bodies, 149; on precession
of the equinoxes, 153; Principia, 147, 150;

on prisca sapientia, 164; on prisca theologia,
160; on resistance of medium, 156; student
notebook of, 154; on theology, 147, 149, 158,
160, 162; on tides, 153; on the Trinity, 160;
on uniformity of nature, 152; on universal
gravitation, 153; on vortices, 152
Newtonianism: and calculus, 151; Voltaire on,
139, 164
nomenclature: in anatomy, 36; in natural
history, 133, 135–6

observation, 144; by alchemists, 118; in
astronomy, 165; by Babylonian astronomers, 13; in natural history, 133–4, 165; by
Ray, 139; by Tycho, 52–3
occult qualities: Descartes on, 87–8; Fludd
on, 122; Gassendi on, 86
Ockham, William of, on divine will, 90
Oldenburg, Henry, 149, 167
optics, 111; Alhazen on, 103; Descartes on,
104, 108, 111; Greeks on, 102–3; Kepler on,
104; as mixed mathematics, 103; Newton
on, 113–7, 149
Oresme, Nicole, 12
Osiander, Andreas, 52; and preface to De
revolutionibus, 44

Paracelsianism: in Becher, 130; in Fludd, 121;
in Van Helmont, 174
Paracelsus (Philippus Aureolus Theophrastus
Bombastus von Hohenheim): on alchemy,
118, 120; on astrology, 120; on chemical
philosophy, 119; on disease, 120; on experiments, use of, 120; on Galen, 119; Hermetic
cosmology in, 118–20; on matter, 120; on
medicine, 118–9; on syphilis, 120; teaching
style of, 119; on weapon salve, 121
parallax, annual stellar, 51–2
Pardies, Ignace-Gaston, criticism of Newton's
theory of colors by, 117
Pascal, Blaise, on the barometer, 79–80
patronage: and Galileo, 1, 63; and natural history, 136; of science, 166
Paul V (pope), 69
Pecham, John, 103
Pellet, Thomas S., and Newton's manuscripts,
147
Peurbach, Georg, 42–3
philosophers' stone: Boyle on, 92; Geber on,
23

*Philosophical Transactions of the Royal Society of London*, 114, 149, 167
phlogiston, 130
*Physiologus*, 25
physiology, 132, 165; Galen on, 28, 141; Harvey on, 141
Pico, Giovanni della Mirandola, on astrology, 71–2
pineal gland, Descartes on, 109
planetary motions, 15; Copernicus on, 45–9; Kepler on, 54–5, 57–9; Tycho on, 53
planets: periods of, 48; stations and retrogradations of, 15, 47
plastic spirit, Kircher on, 137
Plato, 31–2; Newton on, 164
Platonic solids, Kepler on, 54–6
Pliny, 32; on beavers, 24–6; correction of, in the Renaissance, 133; on natural history, 24–5, 132
Polo, Marco, 41
Pompanzzi, Pietro, 144
Pontano, Giovanni, on astrology, 73
precession of the equinoxes, Newton on, 153
pre-Socratic philosophers, Newton on, 164
Priestley, Joseph, on chemistry, 130
principle of accommodation: Augustine on, 67; Galileo on, 66–7
principle of least action, 113
principle of sufficient reason, 162
printing, 32; and illustrations in anatomy, 34–6; and illustrations in natural history, 133, 135–6
*prisca sapientia* (ancient wisdom), Newton on, 164
*prisca theologia* (ancient theology), Newton on, 160
projectile motion, Galileo on, 98–9
providence: Epicurus' denial of, 77; Gassendi on, 78, 89–90; Kepler on, 54; Leibniz on, 164; and the mechanical philosophy, 89
Ptolemy, Claudius, 33: Arabic criticism of, 18; on astrology, 17; on astronomy, 3, 15–19; Averroes' criticism of, 42–3; on geography, 40–1; on vision, 103
pulmonary circulation: Colombo on, 140–1; Harvey on, 141; Ibn al-Nafis on, 140
Pythagoreanism, 32; in Fludd, 121; in Kepler, 55; Newton on, 164

qualities: Aristotelians on, 84; Boyle on, 129; Descartes on, 84, 87; Gassendi on, 84–5, 87; occult, 86–8, 122
Qur'ān, Averroes on, 11

rainbow: Alhazen on, 110–1; Aristotle on, 110; Descartes on, 109–12; experimental model of, 111; geometry of, 110–1; Theodoric of Freiborg on, 110; Witelo on, 110–1
Ray, John, on natural history, 139–40
reflection of light, law of, 106
refraction of light: Descartes on, 112; Fermat on, 113; Huygens on, 113, 115; index of, 115; law of, 113, 115; Newton on, 115
Regiomontanus, Johannes, 43
Renaissance humanism, 30–1, 33–4, 165; and natural history, 133; and Pliny, 133
respiration, experiments on, 143–4
retinal image: Descartes on, 109–10; Kepler on, 104
Rhazes, on alchemy, 21
Rheticus, Georg Joachim, 44
Ricci, Ostilio, 96
Roberval, Gilles Personne de, on the barometer, 80
Royal Society of London, 91, 114, 143, 149; Bacon's influence on, 137; and chemical philosophy, 124; and natural history, 137
Rupecissa, John of, on alchemy, 23

science: definition of, 166; modern, 166; of motion, 94–103, 117, 165; and scripture, 39–40
scientific journals, 114, 167
Scientific Revolution, 167
scientist, definition of, 167
seminal principles, Van Helmont on, 125
Sennert, Daniel, on the weapon salve, 123
sensation, Descartes on, 108–9
Sextus Empiricus, 38
signatures, doctrine of, 32
skepticism: Descartes on, 82; and natural philosophy, 39; and rule of faith, 39; Sextus on, 38
Society of Jesus, 38. *See also* Jesuits
soul: in animals, 144–5; Aristotle on, 9–10, 145; Averroes on, 11; Boyle on, 92, 145; Descartes on, 145; Gassendi on, 80, 145; immortality of, 144–5; Pompanzzi on, 144
space, 100

species: and biblical account, 138–9; extinction of, 139; and fossils, 138–9
Stahl, George Ernst, and Boyle, 130
Starkey, George: and alchemy, 126; and Boyle, 126; as Eirenaeus Philalethes, 126
stations and retrogradations, 47
Steno, on fossils, 137–8
Stenson, Niels, 137–8
substantial forms: in Aristotelianism, 84–5; the mechanical philosophers on, 85
Sylvius, Jacobus, 34
syphilis, Paracelsus on, 120

Tartaglia, Niccolò, 96
teleology: in Aristotelian natural history, 23–4; Aristotle on, 7–8; in Galen, 27, 35, 144
telescope, the, 61–4; Descartes' explanation of, 109; Galileo's justification of, 63–4; Jesuits and, 64; Kepler on, 104
Theodoric of Freiborg, on the rainbow, 110–1
Theophrastus, on natural history, 24, 136
tides: Galileo on, 70; Newton on, 153
Topsell, Edward, 137–8
Torricelli, Evangelista: on the barometer, 79; on impact, 101
translations, 30; of Arabic into Latin, 4, 11; of the Bible, 37; of Greek into Arabic, 4, 11; of Ptolemy, 42
transmutation, 127; in Alexandria, 24; Avicenna on, 22; Rhazes on, 22; Van Helmont on, 126
transubstatiation, 12
Trinity, the, Newton on, 159–60
Tūsī couple, 19; use of, by Copernicus, 43
two-book metaphor, 39, 67
Tycho Brahe, 60, 166; on astrology, 74; on comets, 53; and geo-heliocentric astronomy, 53; and macrocosm and microcosm, 74; observations of, 52–3

uniform circular motion, 42–3, 54, 107; Copernicus on, 43; in Greek astronomy, 13, 15; Kepler's rejection of, 55, 59

uniformity of nature: Galileo on, 65; Newton on, 152
universities, 4–5; chemical philosophy on, 119; and medical gardens, 133–4; and natural philosophy, 167
Urban VIII (pope), 69–70; and Galileo, 70

valves, cardiovascular, 141
Van Helmont, Joan Baptista: on alchemy, 125–6; and Bible, 124–5; and experimental methods, 124–5; on God, 124–5; influence of, on Boyle, 126; and Paracelsus, 124; and phlogiston, 130; theology in, 124–5; on transmutation, 126; and willow tree experiment, 126
Vasco da Gama, 40–1
Venus, phases of, 62
Vesalius, Andreas, 34–6; and Galen's downfall, 144
Vespucci, Amerigo, 41
vision: Alhazen on, 103; Descartes on, 108; Greeks on, 103; Kepler on, 104; science of, 103–117
vitalism, in Van Helmont, 126
Vittore, Fausto, 95
void: Aristotle on, 8; Boyle on, 129; Descartes' denial of, 82; Epicurus on, 77, 79; Galileo on, 97; Gassendi on, 78; Huygens on, 101; Lucretius on, 78
Voltaire, on fossils, 139
vortices: Descartes on, 83, 112; refutation of, by Newton, 152

wave-theory of light, Huygens on, 113–5
weapon salve, 121–4
will, divine: Aquinas on, 90; Descartes on, 91; Gassendi on, 90–1; Leibniz on, 163–4; Newton on, 163–4; Ockham on, 90
Witelo, Erazmus Ciolek, 103, 110–1
world, eternity of, 2; in Aristotle, 5; Averroes on, 11

Zosimos of Panopolis, 21